JEFFERSON'S GAS WELDING MANUAL

by
Ted B. Jefferson

CONTENTS

Preface

There are many books on welding, all serving a purpose, unfortunately, most of these books are designed to teach a person how to weld. In attempting to carry out this aim, it is found that much of the subject matter is the same, but only told in different words. After twenty-five years of reading welding books which were repeating themselves it seemed time to bring out a new kind of welding book.

In this volume the author assumes that the reader has learned to weld but that occasional helps are needed. Those who do not know how to gas weld or flame cut will find that the fundamentals have been included in this volume so that they will be able to do these jobs. The experienced weldor and the advanced student will find this volume to be a means of increasing their proficiency.

The most valuable data is presented in Chapter 2. A series of *Jiffy Welding Guides* present concisely and precisely the basic data required for the oxyacetylene welding of any metal weldable by this process. A concise chart in Chapter 3 quickly aids in the selection of the proper means for the joining of dissimilar metals by the oxyacetylene process.

Flame cutting is also covered without frills. Again the reader is shown the best flame cutting technique for the most efficient serving of metals. While confined to manual flame cutting, the Guides cover the flame cutting of all types of ferrous metals.

In other tables and charts the author has tried to compile data to make this a most complete and compact book on the oxyacetylene welding and cutting process. To those in the welding industry whose thoughts, ideas and suggestions caused this book to come into being, the author extends his appreciation.

Skokie, Illinois, December 1, 1951.   *Ted B. Jefferson*

CHAPTER 1

How To Weld Metals

Any metal may be welded provided the proper technique and the correct welding process is used. The statement, with some modifications, applies to oxyacetylene welding processes. Practically every metal may be welded by this welding process, though a few tricky materials may possibly be welded more satisfactorily by some other method.

Be Sure the Metal Is Clean. Too much emphasis cannot be placed upon cleanliness when it comes to welding. Dirt in a joint can cause a lot of trouble and will lead to such defects as: incomplete welds, gas pockets, blowholes and inclusions. While some of these defects may not seem important it is never known when they may prove to be quite serious. The weldor should always strive for the perfect weld!

Oil and grease may be readily removed from the base metal using a caustic soda and water solution or carbon tetrachloride. Gasoline is an excellent solvent, but the fire danger is too great to recommend its use. Remember, when using gasoline, the something that burns may be you. Flame-cleaning may also be used to remove oil and grease as well as paint.

Shop dirt and light rust on the base metal may be removed by chipping and wire brushing. Heavy rust and scale should be removed by manual or mechanical chipping, flame-cleaning, or grinding and wire brushing.

1

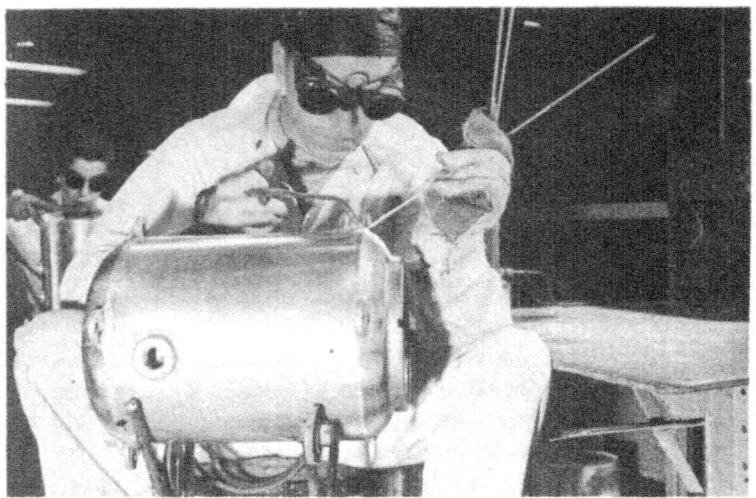

Fig. 1-1. Oxyacetylene welding often provides the most satisfactory solution to a welding problem.

The light metals, such as aluminum and magnesium, should be thoroughly cleaned with a chemical solvent (carbon tetrachloride) then the edges should be filed or wire brushed vigorously to remove oxides. Sometimes an acid pickle bath is used for removing these oxides.

In any event, the metals to be welded must be as thoroughly cleaned, in the welding area, as circumstances will permit. Weldors who have become accustomed to welding steel often find it easy to work out, of the weld, the slag resulting from dirty welding conditions. This may be done when welding steel but it is a bad habit. Weldors who do it are apt to depend upon this ability too much when attempting to weld other metals.

Remember, you cannot depend on puddling impurities out of molten metal when you are brazing, braze-welding or hard-facing. Clean surfaces are extremely important in all welding operations. *Be sure you are working with clean base metal.*

2

HOW TO WELD METALS

Types of Weld Joints. Four types of weld joints are used in oxyacetylene welding, they are: butt, fillet, lap and flange joint. While there are many variations of these joints their basic types are shown in Figs. 1-2 to 1-5.

Fig. 1-2. A butt joint should be used wherever possible.

Butt Joint. The butt weld (Fig. 1-2) is the most common and simplest type of weld joint. A butt weld should be used wherever possible for, when properly made, it is the strongest of weld joints. A butt weld is made by welding along the joint formed when two plates are brought together. The manner in which the edges are prepared for welding depends upon the thickness of the plate as shown in Fig. 1-7.

Fig. 1-3. The fillet weld is extremely popular.

Fillet Joint. The second most popular weld joint is that formed by the fillet weld, Fig. 1-3. To get the maximum strength from this type of weld, it is necessary to carry a wide weld pool. The cross-section of the weld pool must be equal to the thickness of the base metal at all points along the seam. Since such a wide weld pool is seldom used, fillet welds should be avoided whenever another type of weld joint can be used.

3

Lap Joint. A lap weld (Fig. 1-4) is used to join the edge of one plate to the face of another, when the plates to be joined are placed one above another. This type of weld is not too satisfactory because it has little resistance to bending. A double lap weld is better from this standpoint but requires twice as much welding and weld metal as butt join.

Fig. 1-4. There is a little resistance to bending in the lap joint unless a double lap weld (right) is used.

Flange Joint. Filler metal is not required for the flange weld, Fig. 1-5. This joint is used on thin gage sheet metal, generally 20 gage or thinner. The height on the flange should be equal to the thickness of the sheet. Welding is done by melting down the flange. It is important that penetration be complete in making this type of weld.

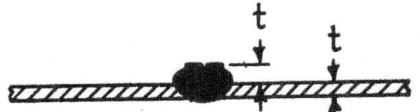

Fig. 1-5. Flange joints are used on thin sheets.

Tack Welds. Tack welds (Fig 1-6) are sometimes used to hold the edges of two pieces in alignment prior to starting a weld. Be sure to melt the tack weld into the weld pool during the welding operation.

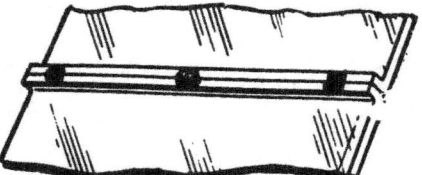

Fig. 1-6. Tack welds aid in alignment.

4

HOW TO WELD METALS

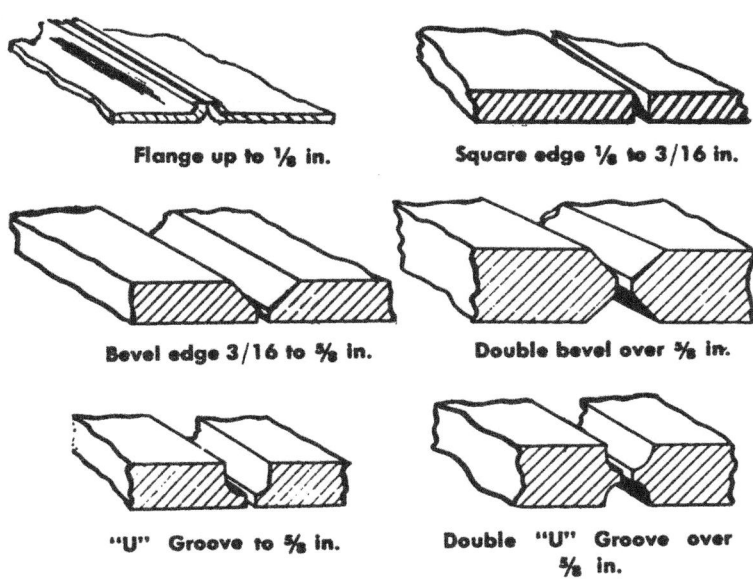

Flange up to ⅛ in.

Square edge ⅛ to 3/16 in.

Bevel edge 3/16 to ⅝ in.

Double bevel over ⅝ in.

"U" Groove to ⅝ in.

Double "U" Groove over ⅝ in.

Fig. 1-7. Different thickness steel plates require different edge preparation.

Fig. 1-8. Notching is a part of the edge preparation of the light metals, aluminum and magnesium.

5

Joint Preparation. The type of weld joint to be made determines the joint preparation to be undertaken. Proper joint preparation is extremely important if good welds are to be produced. The manner in which joints are to be prepared for various thicknesses of low carbon steel are shown in Fig. 1-7. These suggestions will work equally as well for all weldable metals except the light metals, aluminum and magnesium, which should, in addition, be notched, Fig. 1-8.

For a flanged joint the only preparation necessary is the bending of flanges along the edges to be welded.

On square edge butt joints the edges must be accurately matched for welding. If trimming is necessary use an oxyacetylene cutting torch or a mechanical means. Be sure to clean off the slag if the plate were flame cut.

Bevel joints may be prepared by machining, chipping, grinding or flame cutting; whichever is the most convenient and economical. U-grooves may be made easily by flame cutting, using a special cutting tip.

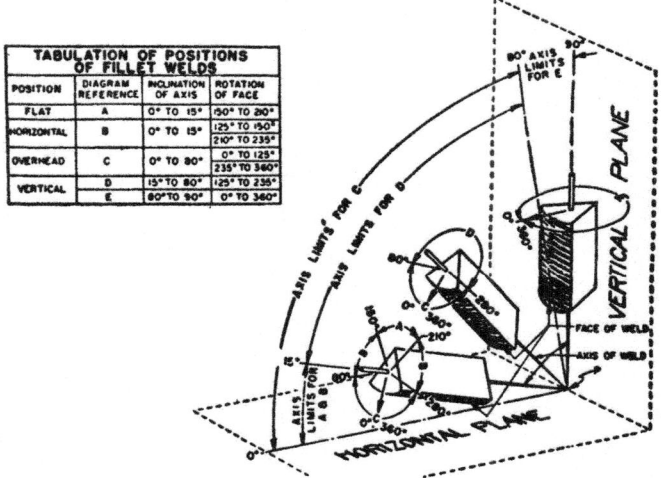

Fig. 1-9. There are four positions of welding.

6

HOW TO WELD METALS

Positions of Welding. All welds may be made in one or more of the four common welding positions. They are: flat or downhand; vertical; horizontal or overhead (Fig. 1-9).

Steel may be readily welded in any of these four positions. This is also true of most other metals though the vertical and overhead positions are extremely difficult with some metals.

There are four factors which play important parts in making position welds, they are: cohesiveness of the weld metal, rod manipulation; flame force and base and weld metal support. The relative importance of these factors for the four weld positions are shown in Table 1-1.

Table 1-1—Weld Position Factors

Factor	Flat	Vertical	Horizontal	Overhead
Support	1	3	2	4
Manipulation	2	1	1	2
Cohesiveness	3	4	2	1
Flame Force	4	2	3	3

Note: 1 is most important factor while 4 is least important

Weld Rods. Filler metal obtained from a welding rod is generally added to the weld puddle in the making of every weld. For satisfactory results, it is essential that this filler metal be similar in metallurgical and physical characteristics to the base metals. Since commercial welding rods are made to definite specifications, it is recommended that they be used rather than "any old wire"; otherwise a good weld may not be obtained.

Table 1-2 gives data on the number of various size welding rods of steel, bronze or aluminum in a 10 pound bundle.

Welding Fluxes. Many metals, which may be readily welded, require a flux to insure a good weld. A flux serves

7

Table 1-2—Gas Welding Rod Data
(Approximate Number of 36 in. rods in 10 lb bundle.)

Size, in.	Steel	Bronze	Aluminum
$\frac{1}{16}$	320	298	800
$\frac{3}{32}$	142	130	356
$\frac{1}{8}$	80	70	240
$\frac{5}{32}$	50	45	125
$\frac{3}{16}$	35	32	88
$\frac{1}{4}$	20	18	50
$\frac{5}{16}$	13	12	33
$\frac{3}{8}$	9	7	24

primarily to prevent oxidation, dissolve oxides, release trapped gases and slag and to cleanse the base metal during welding, soldering and brazing. Flux is used in the powdered form by dipping the heated welding rod in the powdered flux, or in the paste form by painting the flux paste in the welding rod and the joint edges prior to welding. A flux paste is made by dissolving powder flux in water or alcohol.

It is extremely important that a flux be used in the welding of cast irons and most of the nonferrous metals. Upon completion of welds remove all flux and slag.

Tip Size. It is extremely important that the proper size welding tip is fitted in the torch to be used in making a weld on a given thickness of metal. Too small a tip means that it will either take too long to bring the base metal to a welding temperature or that it might even be impossible to make a weld. Too large a tip will probably result in the burning of the base metal.

The recommendations of the welding torch manufacturer as to the proper size tip to be used for a given metal thickness should be closely followed. In the event such recommendations are not available Table 1-3 indicates the tip size and gas pressures, for a balanced pressure torch, when welding various thicknesses of steel.

Table 1-3—Tip Sizes for Welding Steel

Base Metal Thickness	Tip Drill Size No.*	Oxygen, psi	Acetylene, psi
24 ga. to $\frac{1}{16}$ in.	70	3-4	3-4
18 ga.—$\frac{1}{8}$ in.	62	4-5	4-5
$\frac{1}{8}$—$\frac{3}{16}$ in.	54	4-5	4-5
$\frac{3}{16}$—$\frac{3}{8}$ in.	48	5-6	5-7
$\frac{3}{8}$—$\frac{1}{2}$ in.	42	7-9	8-10
$\frac{1}{2}$—$\frac{3}{4}$ in.	37	7-9	8-10
$\frac{5}{8}$—1 in.	33	8-10	9-11
1 in. and up	29	9-11	10-12

While this tip size is recommended, the size of tip either a size large or a size smaller may be used by adjusting the pressures of the welding gases. See Chapter 7 for comparison of various types of welding tips.

Flame Adjustment. An oxyacetylene flame may be one of three types: neutral, carburizing or oxidizing, with the neutral flame generally preferred. These flames are illustrated by sketches in Fig. 1-10.

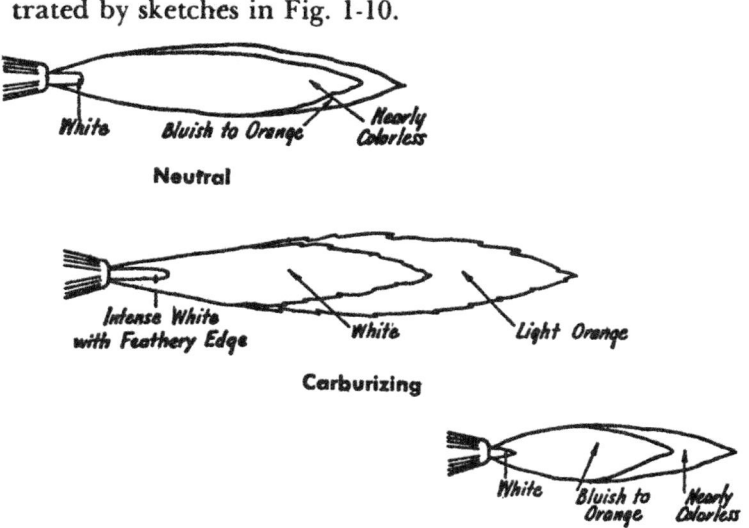

Fig. 1-10. An oxyacetylene flame is one of three types.

9

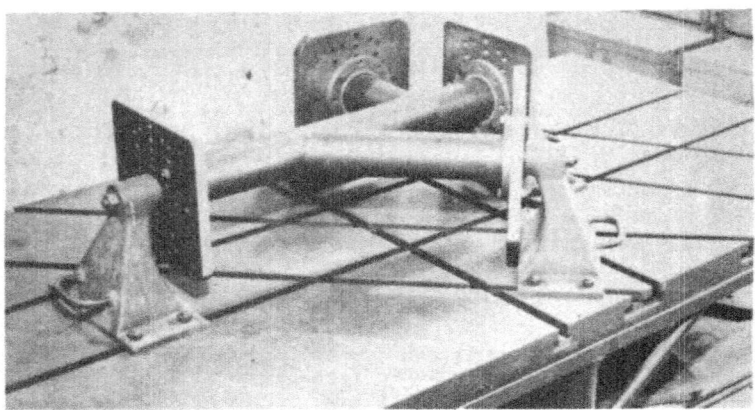

Fig. 1-11. Jigs and fixtures simplify the welding job.

Neutral Flame. The neutral flame is produced by burning the correct mixture of oxygen and acetylene. A clear, well-defined white cone indicates that the flame is correct and that neither gas is being wasted. Welds made with the neutral flame should be free of burned metal or hard-spots and be thoroughly fused.

Carburizing Flame. A carburizing flame is burning an excess of acetylene. It introduces carbon into the weld and may be readily recognized by the feathery edge of the white cone. Some metals are welded with a carburizing flame because it adds carbon to the weld.

Oxidizing Flame. The shorter envelope and the small pointed white cone identifies the oxidizing flame with its excess of oxygen. This flame causes an oxidizing or burning of the weld metal.

Jigs and Fixtures. The appearance of the finished welded product is often dependent upon the manner of alignment of the prepared edges to be welded. The degree of care exercised in aligning parts will vary with different

10

jobs and different materials. For example, while steel plate may be welded without the use of jigs, aluminum pieces generally always require jigging.

Jigs and fixtures are nearly essential, if it is necessary that the parts or surfaces line-up accurately when welding is completed. Without their use it would probably be necessary to spend considerable time in finishing the weld. Some jigs are simple affairs such as that used in aligning the pipes in Fig. 1-11. An even slightly less complicated fixture, Fig. 1-12, is used for aligning the corners of small frames on a production welding job.

Fig. 1-12. This jig guarantees that all pieces will be the same.

When welding some materials, especially thin gage aluminum it is desirable to dissipate the welding heat quickly. This is done by using a fixture fitted with chill blocks (Fig. 1-13). The grooved back-up bar of this fixture helps control the shape of the underside of the weld bead.

Tack welds are also used to maintain alignment between parts being welded.

Jigs, in addition, to holding parts in their correct position are used to speed welding operations where many

11

pieces of the same kind are to be fabricated, Fig. 1-13.

Carbon blocks, rods and carbon composition or paste are frequently used as a substitute for jigs and fixtures in the alignment of parts. This is especially true when making repairs on castings since carbon sheets or blocks are available in many sizes and may be readily shaped for the job at hand.

The use of carbon blocks in conjunction with vertical welds of cast iron make it possible to perform the heaviest of welds on the casting without dismantling them. It is also possible to reinforce the weld to any thickness desired by using carbon blocks than the wider weld and grooving the blocks so as to make the thicker weld possible.

Carbon paste is an excellent alignment aid inasmuch as balls of paste can be used to provide a cushion under the piece being welded. The weldor knows that the piece is then supported uniformily and there is no chance for sagging.

There are many other possible applications for carbon paste and blocks. Examples of a few of these uses are shown in Fig. 1-14.

Preheating. Preheating of the base metal prior to welding or flame cutting operations is quite common. Preheating offers the advantages of: (1) reducing welding time; (2) reduces or prevents shrinkage stresses; (3) re-

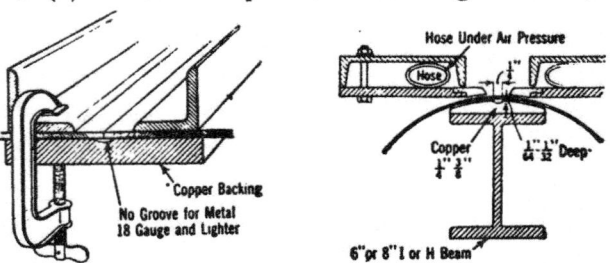

Fig. 1-13. Jigs with back-up bars.

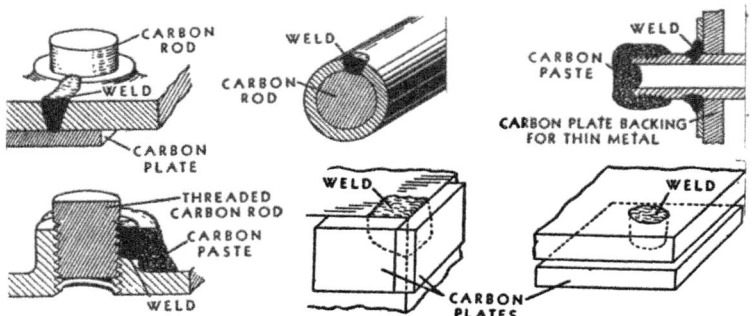

Fig. 1-14. Carbon rods, plates and paste are welding aids.

duces distortion; (4) eliminates the danger of crack formation and reduces the hardness of the weld deposit.

The method of preheating depends upon the size, shape and material of the piece to be welded. In some instances, especially with small pieces or where localized heating is satisfactory, a welding torch may be used for preheating. Such an application is shown in Fig. 1-15. On other occasions a specially designed preheating torch, burning city gas, LP gas or oil, is used, Fig. 1-16, or a fire brick furnace, Fig. 1-17.

Fig. 1-15. Another welding torch often provides the preheat.

13

A preheating furnace is the most common of preheating devices. While gas or oil may be used as a fuel for preheating, charcoal is the most popular fuel.

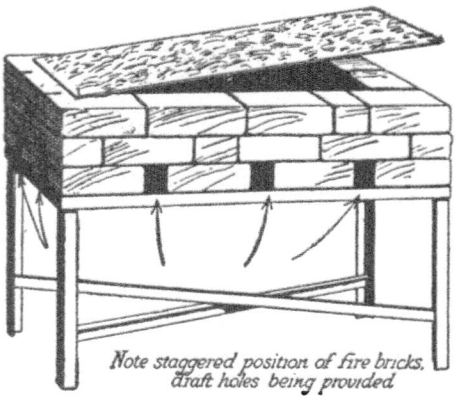

Note staggered position of fire bricks, draft holes being provided

Fig. 1-16. A special torch designed for preheating is highly efficient.

Fig. 1-17. Large castings are preheated in fire brick furnaces.

14

Since preheating should be carried on during welding, the furnace should be constructed so as to allow sections to be opened without destroying the fire. A good furnace bottom is essential and should be built of firebrick. The walls of the furnace should also be of firebrick and should be spaced so as to provide plenty of room around the piece to be preheated. Draft holes should be left around the bottom of the wall to insure uniform burning. Charcoal will burn without a draft and with a draft will develop a white heat. For the latter reason, care should be taken not to place draft holes near thin sections of the piece being preheated. Wherever a draft hole is too large it may be blocked off.

The roof of the furnace should be made of a layer or two of sheet asbestos supported by steel rod where necessary. A number of holes should be punched in the asbestos to allow the charcoal fume to escape. Because of these fumes the preheating should be carried on in well ventilated areas. When the preheated part is ready for welding it is a simple matter to tear open holes in the asbestos to permit access to the break, Fig. 1-18.

Fig. 1-18. Welding is done through access holes in the furnace.

15

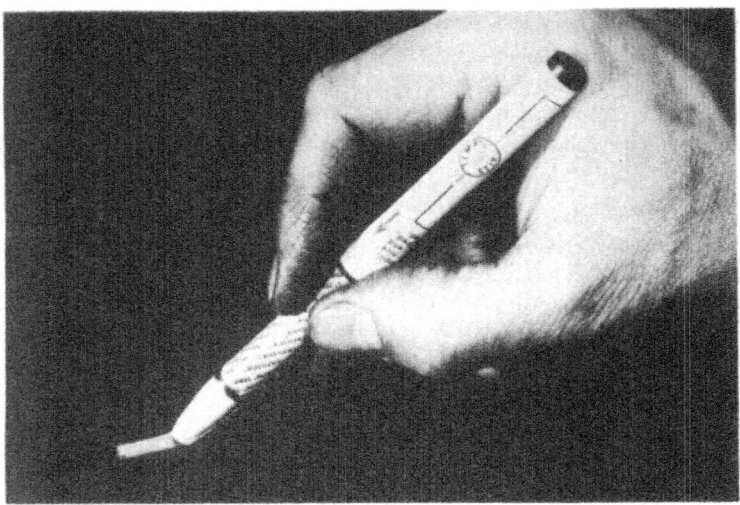

Fig. 1-19. A glistening liquid streak indicates the piece to be at or above preheat temperature while the dry chalky mark shows the area is below temperature.

Preheating Temperatures. For most materials it is quite essential that they be preheated to a recommended temperature and maintained at that temperature during the entire welding operation. Cast iron may be taken up to red heat, 1000-1500 F, without serious effects while aluminum must not be heated over 800 F. Generally, care must be taken not to overheat metals during preheating.

For ferrous materials (iron and steel) it is possible to determine their approximate Fahrenheit (F) temperatures by their color changes during heating. Table 1-4 gives the approximate temperature for the various changes in color during the heating of steel. Non-ferrous materials, however give no color-change indications and their temperature must be determined in some other manner.

The most convenient and accurate means of determining steel temperatures in the black heat ranges (below 900 F) and for those metals which do not indicate their tempera-

16

ture by color change, is by use of a patented temperature-indicating crayon, pellet or paint. The temperature-indicating crayons, tempilsticks, are available over a wide range of temperatures. Tempilsticks are available which melt at 12½ degree intervals between 113 F and 400 F and at 50 degree intervals from 400 F to 2000 F.

Table 1-4—Temperature—Color Chart

Color	Temperature, F
Brilliant White	2732
White Heat	2552
Yellow-white	2372
Orange-yellow	2192
Orange-red	2012
Bright Cherry Red	1832
Cherry Red	1652
Dull Cherry Red	1472
Dark Red	1292
Red in sunlight	1077
Red in daylight	975
Faint Red in twilight	885
Faint Red in dark	752

All that is involved in the use of a tempilstick is to select one which melts at the desired temperature and mark the piece to be heated to that temperature. Now heat the piece uniformly until the crayon mark melts, Fig. 19, to indicate that the desired temperature has been reached. Tempilsticks are guaranteed to be accurate within 1%, that would be within 4 degrees for a 400 F tempilstick.

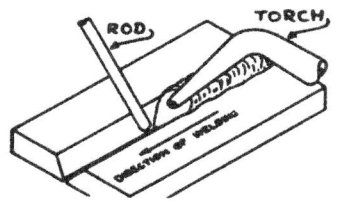

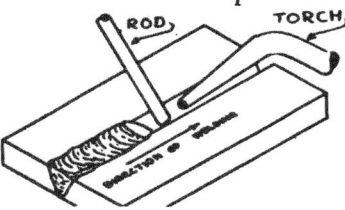

Fig. 1-20. Forehand welding.   Fig. 1-21. Backhand welding.

17

Welding Techniques. Two common techniques of welding are employed in oxyacetylene welding—forehand welding and backhand welding. Become familiar with both of these techniques so that they might be used to their best advantage.

Forehand Welding. In forehand welding the torch tip is inclined slightly forward in the direction of the weld and the welding rod is held ahead of the flame, Fig. 1-20. It is evident that, with the flame and rod in these positions, the rod will prevent the heating zone of the flame from reaching the bottom of the joint ahead of the weld puddle.

This is advantageous on thin material because it prevents melting to such an extent as to cause a large hole. But these relative positions of the torch, welding rod and flame has a distinct disadvantage on thick materials. In fact, when forehand welding is attempted on thick material an oxidized weld is the usual result. To reach the bottom of the groove with the flame, the rod must be pulled out of the puddle and the protective envelope of the flame. By this action the rod becomes oxidized from contact with the air. For this reason, forehand welding is generally confined to the welding of thin sections.

In forehand welding, the technique requires that the welding rod be moved slightly to one side allowing the flame to reach the bottom in one direction, then with a semi-circular movement the positions of the rod and the flame are reversed to heat the other side. During this movement care must be exercised to avoid contact between the tip of the white cone and the end of the rod.

It will be found easier to remember this system if the movement of the torch and rod are practiced separately at first.

Watch These Steps. In forehand welding there are five points that should be checked to insure satisfactory welds·

(1)   The white cone must be kept from touching the welding rod;

(2)   Heat must be allowed to penetrate to the very bottom of the joint for the entire length of the weld;

(3)   The sides of the weld must be practically parallel to insure a uniform deposit;

(4)   Enough weld metal must be added to form a full bead and

(5)   Weld metal must be evenly distributed.

Backhand Welding.   In backhand welding the flame is inclined backward (Fig. 1-21) upon the weld metal that has been deposited.   The flame is given a slight transverse movement while the welding rod, following the flame, is given a more pronounced transverse, semi-circular movement.   The tip of the torch is held so that it makes about a 70 degree angle with the plane of the weld.   The angle between the rod and the plane of the base metal is usually about 45 degrees.

Because of this rod angle, welds produced, using the backhand technique, are rippled in character.   This is caused by the manipulation of the rod in the weld puddle.

Backhand Technique.   In backhand welding, the welding rod does not prevent the flame from reaching the bottom of the groove or joint.   The welding rod is manipulated in a manner which will thoroughly mix the molten metal in the weld pool.   At the same time the movement keeps the rod away from the flame to insure thorough melting of the top portion of each ripple.

Backhand Welding Advantages.   Backhand welding is especially desirable when welding any metal over 1/4 in. thick.   The big advantage is good penetration since the rod is not a handicap and good fusion results due to constant weld puddle agitation.   Backhand welding is also faster because of the preheating action of the torch flame.

The direction of the flame prevents oxidation since it covers the weld pool. It is also recommended that a narrower bevel (60 degree) be used on backhand welding; a factor which reduces costs and speeds up production.

To obtain the fullest benefits of these advantages be sure to use the proper size welding tips and correct gas pressures as well as the proper diameter welding rod. The size of the welding rod is extremely important; too small a rod melts too rapidly making it difficult to properly distribute the weld metal. Too large a welding rod slows down melting and reduces the speed of welding.

Welding Pointers. There are certain principles involved in welding operations which should be reviewed prior to undertaking any welding job.

● Always be sure that the parent metal along the line of welding is heated to the proper melting point.

● Be sure to melt the base metal on *both* sides of the joint the entire length of the weld.

● Heat the welding rod and the base metal to a melting temperature before starting to weld.

● Puddle the base metal and the filler metal together properly.

● Filler metal from the welding rod should be added as steadily and as evenly as possible.

● Use the outer envelope of the oxyacetylene flame to protect the weld metal from oxidation.

● Don't reheat weld metal which has cooled.

Remember there are many methods by which the torch and the welding rod may be manipulated during a welding operation. Be sure that your movements are systematic and properly executed. Should you desire to change your technique, resist the impulse until the weld you are working on has been completed.

Be sure that the welding rod is never melted above the weld pool. Do not allow molten weld metal to drop into the weld puddle a drop at a time. When the welding rod begins to melt and fills the groove advance the flame slowly along the line of welding. The end of the rod should be kept in the weld puddle while the pool of molten metal makes steady progress in the required direction of welding.

Steel Welding Fundamentals. While it is known that most metals may be welded by the oxyacetylene process, most of the welding by this process is on steel. For this reason, there is included in the chapter many fundamentals dealing with the welding of steel which should be remembered. Most of the facts applying to steel apply equally as well to other materials. This is clearly shown by *"Jefferson's Jiffy Welding Guides,"* Chapter 2.

Remember these steel welding fundamentals:
- Maintain a neutral flame at all times. It is essential for satisfactory steel welding.
- Watch out! Steel warps and deforms when heated. Avoid excessive heating for it is expensive.
- A flux for dissolving scale or oxide is not necessary in steel welding.
- Too large a flame will overheat the base metal and cause it to spark and burn.
- Keep the protective envelope of the flame over the weld puddle.

Exposure of the weld puddle to either oxygen or acetylene will result in oxidizing or carburizing the weld. Removal of the flame from the weld puddle causes oxidation since the molten steel combines with the oxygen in the air. For the same reason, the end of the welding rod should be kept in the melted weld pool or the flame envelope.

Always remember—the *skill of the weldor* and the *quality of the filler metal* determines the *strength of the weld joint.*

Cleaning the Weld. All slag, oxide or flux must be removed before a welded piece may be satisfactorily painted. Make it a practice to chip and wire brush each

21

Characteristics of Sparks Gener-

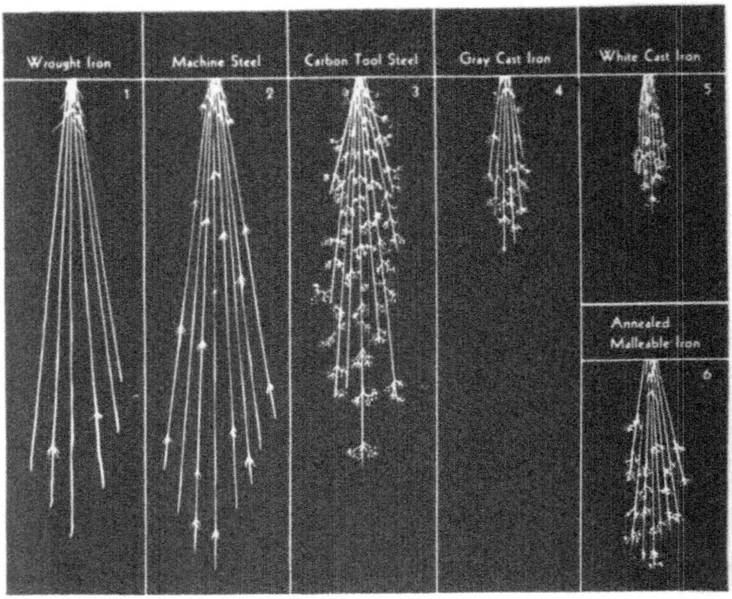

	Metal	Volume of Stream	Relative Length of Stream, Inches†
1.	Wrought iron	Large	65
2.	Machine steel	Large	70
3.	Carbon tool steel	Moderately large	55
4.	Gray cast iron	Small	25
5.	White cast iron	Very small	20
6.	Annealed mall. iron	Moderate	30
7.	High speed steel	Small	60
8.	Manganese steel	Moderately large	45
9.	Stainless steel	Moderate	50
10.	Tungsten-chromium die steel	Small	35
11.	Nitrided Nitralloy	Large (curved)	55
12.	Stellite	Very small	10
13.	Cemented tungsten carbide	Extremely small	2
14.	Nickel	Very small**	10
15.	Copper, brass, aluminum	None	

†Figures obtained with 12″ wheel on bench stand and are relative only. Actual length in each instance will vary with grinding wheel, pressure, etc.

Fig. 1-22.   The spark test provides a

- - - -

22

ated by the Grinding of Metals

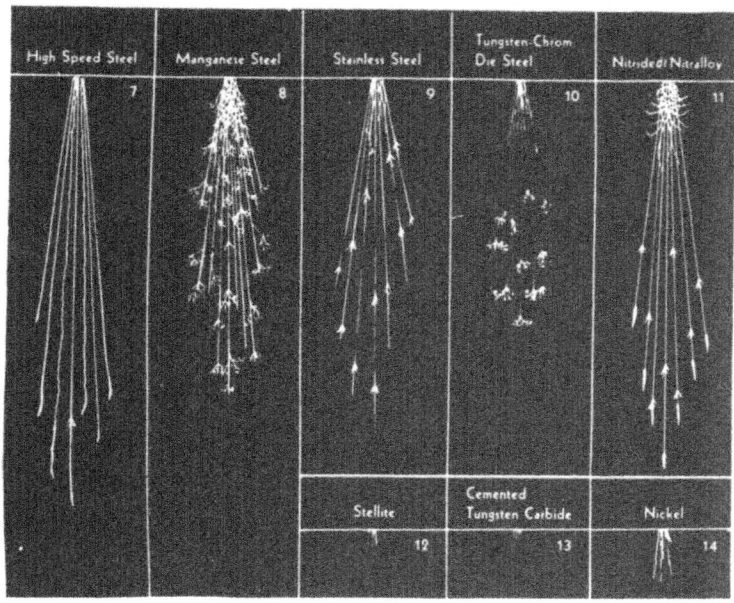

Color of Stream Close to Wheel	Color of Streaks Near End of Stream	Quantity of Spurts	Nature of Spurts	
Straw	White	Very few	Forked	1.
White	White	Few	Forked	2.
White	White	Very many	Fine, repeating	3.
Red	Straw	Many	Fine, repeating	4.
Red	Straw	Few	Fine, repeating	5.
Red	Straw	Many	Fine, repeating	6.
Red	Straw	Extremely few	Forked	7.
White	White	Many	Fine, repeating	8.
Straw	White	Moderate	Forked	9.
Red	Straw*	Many	Fine, repeating	10.
White	White	Moderate	Forked	11.
Orange	Orange	None		12.
Light Orange	Light Orange	None		13.
Orange	Orange	None		14.
		None		15.

*Blue-white spurts. **Some wavy streaks.
(Reprinted from (Norton Co.) "Grits and Grinds" June 1940)

simple means for identifying metals.

23

Table 1-5—Metal Identification Chart

Color	Magnetic Properties	Chip Characteristic	Chip Size	Specific Gravity	Name of Alloy
silvery stainless	non-mag.	smooth easily cut	continuous	8.9	Monel
reddish yellow	non-mag.	smooth easily cut	continuous	8.8 to 9.0	bronze
silver	non-mag.	smooth easily chipped	continuous	8.7 to 8.8	nickel silver
yellow or reddish	non-mag.	smooth easily cut	continuous	8.4 to 8.8	brass
yellow gold	non-mag.		continuous	7.5 to 8.2	aluminum bronze
silver stainless	non-mag.	smooth easily chipped	continuous	7.6 to 7.8	stainless steel
silvery	magnetic	smooth easily chipped[1]	continuous	7.6 to 7.8	steel
silver	magnetic	rough and tough	1/4-3/8 in.	about 7.6	malleable iron
silver	magnetic	smooth but brittle[2]	1/8 in.	7.0 to 7.6	cast iron
silver	non-mag.	smooth brittle		7.0 to 7.6	pewter
silver	non-mag.			6.7 to 6.8	zinc alloy
silver	non-mag.	smooth easily cut	continuous	2.7 to 3.0	aluminum alloy
silver	non-mag.	smooth easily cut	continuous	2.0 to 2.5	aluminum magnesium alloy

1. For low carbon steel. High carbon steel has fine grain fracture and is very hard but may be chipped.
2. Gray cast iron as indicated. White cast iron is very brittle.

24

weld whether it is to be painted or not. This cleaning operation will reduce the tendency toward corrosion and will give a finished appearance to each welding job.

Post Heating. To insure the satisfactory welding of some metals it is necessary that they receive some type of after weld heat treatment. In some cases, this post heating is merely an annealing operation brought about by slowly cooling from the preheating temperature. In other instances, the post heating is a definite heat treating operation for the purpose of stress-relieving, grain refinement or restoring definite physical characteristics.

Post heating recommendations should be carefully followed. Specific temperature recommendations should be observed and accurately determined by tempilsticks.

Metal Identification. The proper welding procedure may be easily planned when it is known just what metal is to be welded. Color will sometimes give a quick clue to the identity of a metal. Occasionally, however, a broken piece requires repairs and from its outward appearance it is impossible to determine just what metal or alloy it may be. There are several tests which will aid in the general identification of metals. With the clues obtained satisfactory repair welds may be made, since the weldor will know the general type of metal he is welding.

Spark Test. The most common test is the spark test. The piece to be identified is touched against a grinding wheel. Identification is made by matching the characteristic sparks produced with those illustrated in Fig. 1-22.

Chip Test. Another relatively simple test is the chip test. In this instance the metal to be identified is chipped with a cold chisel. Identification is made by comparing the size of chips, color of metal, hardness and surface condition of the chipped metal with a known metal. Often this test requires additional tests covering magnetic properties and specific gravity determinations before the metal is really identified. When these three characteristics have been determined, the metal may be identified by use of Table 1-5.

Magnetic Properties. Magnetic properties are, of course, determined with a magnet. If the metal is attracted by a magnet it is magnetic.

Specific Gravity. Specific gravity is somewhat more difficult to determine. Use a spring scale and weigh the object being tested in air and then submerged in water. Determine the difference between the two weights. Specific gravity is then found by the formula:

$$\text{Specific gravity} = \frac{\text{Air weight}}{\text{Air weight minus Water weight}}$$

Hardness Testing. Various steels may be identified by their hardness. Hardness may be approximated by a simple file test. It is done by determining the resistance of the metal to the cutting action of the machinist file. When the surface reaction is compared with the data of Table 1-6, the approximate Brinell or Rockwell hardness number may be determined. The type of steel is also indicated in Table 1-6.

Table 1-6—File Hardness Test*

Approximate Hardness		Surface reaction to filing	Type of
Brinell	Rockwell		steel
100	60B	Metal is readily cut by file.	Low carbon steel
200	15C	Metal is readily cut by file under moderate pressure.	Medium carbon steel
300	30C	Metal difficult to file though it can be cut.	High alloy steel
400	40C	Metal is cut only with greatest of effort.	High carbon steel
500	50C	Metal nearly impossible to cut.	Tool steel
600	60C	Metal cannot be cut with a file.	Hardened tool steel

Characteristics when using new, machinist's hand file.

HOW TO WELD METALS

Safety. Too much emphasis cannot be placed on keeping welding and cutting operations safe. In gas welding and cutting operations the operator is "playing with fire" but the flame is, or should be, under control at all times. There is little danger as long as the operator maintains control of the flame and sees that it is not directed on inflammable material.

The manufacturers of welding and cutting equipment have designed the torches, regulators and other pieces so they are safe so—use them safely.

Oxygen Cylinder Safety. Use only cylinders approved by interstate commerce commission for transportation of compressed gases.

- Don't refer to oxygen as "air".
- Don't permit oil or grease to come in contact with cylinder, valves, regulators, gauges or fittings.
- Don't handle with oily gloves.
- Don't use oxygen from a cylinder without properly reducing regulator.
- Don't allow welding or cutting sparks or flame to contact cylinder.
- Don't try to mix gases in a cylinder.
- Don't hammer on or throw cylinders around.
- Don't use oxygen as a substitute for compressed air.

Acetylene Cylinder Safety. Use only approved acetylene cylinders and keep them in a standing or near vertical position.

- Don't refer to acetylene as "gas"
- Don't use acetylene at a pressure exceeding 15 psi.
- Don't test for acetylene leaks with an open flame. Use soapy water.

General Cylinder Safety. Don't use cylinders as rollers or work supports.

- Don't tamper with cylinder safety valves.
- Don't hang torches on regulators or cylinders.

27

CHAPTER 2

Jiffy Gas Welding Guides

The oxyacteylene welding of any metal, weldable by the process, requires the weldor to perform a number of distinct operations. These operations may be grouped into three steps; (1) preweld preparation; (2) welding and (3) postweld treatment.

Preweld preparation covers everything necessary to prepare a piece for welding. This includes determining: that the piece to be welded is clean; whether joint preparation is required; if the piece must be positioned, placed in jigs or fixtures and if preheating is necessary. Preweld preparation also includes the selection of the proper welding rod and flux for the job to be undertaken.

The actual welding operation is the second step. This involves the selection of the correct flame, its adjustment and the using of the proper welding technique to complete the weld.

Postweld treatment is the third step. This phase of the overall welding operation may be omitted in the case of some metals while for others it is a step of extreme importance. Postweld treatment involves postweld cleaning, post heating and the finishing operations.

The breaking down of a welding operation into three steps is not new. Instead, it is common practice and most weldors follow these steps closely when welding steel. Tests have proved, a weldor familiar with the welding of steel

28

would be able to weld any other weldable metal, if the steps involved could be compared with the steps in welding steel.

The oxyacetylene welding of steel is a comparatively simple task after the fundamentals of welding have been mastered. When the welding characteristics of other metals are compared to those of steel, and the welding technique changed to meet the new conditions created by these new materials, they may likewise be welded easily. The *Jiffy Gas Welding Guides* included in this chapter have been designed for this purpose.

By following the data presented in these Guides it will be possible for the weldor, whether he is comparatively new at the game or an old timer, to make satisfactory oxyacetylene welds on any weldable metal. When the weld is on a piece of new metal there will be no doubt as to which Guide is applicable. When working on old metal, however, care must be exercised to properly identify the material to be sure that the correct welding procedure is employed. If there is any doubt as to the identity of the metal it should be tested as outlined in Table 1-6.

In a concise form the Jiffy Welding Guides present all the essential data for the welding of most of the metals encountered in normal gas welding operations. Two Guides of a general nature (braze-welding and silver brazing) have been included with those covering the fusion welding of specific metals. Since the steps involved in the braze-welding of any metal (which may be joined by this process) are the same, this application is covered by a general chart rather than one covering the braze welding of a specific metal. The same is true of silver brazing applications.

The *Jiffy Gas Welding Guides* have been arranged alphabetically, i.e., Aluminum, Brass, Bronze, etc. Table 2-1 suggests means of using the oxyacetylene flame for joining metals which are not included in the Jiffy Welding

Table 2-1   Oxyacetylene Weldable Metals

Metal	Jiffy Guide
Aluminum 2S	Aluminum 2S
Aluminum 3S	Aluminum 2S
Aluminum 43	Aluminum 2S
Aluminum 52S	Aluminum (H-T)
Aluminum 53S	Aluminum (H-T)
Aluminum 61S	Aluminum (H-T)
Brass, Admiralty*†	Brass
Brass, Naval*†	Brass
Brass, Red*†	Brass
Brass, Yellow*†	Brass
Bronze, Aluminum	Silver Brazing
Bronze, Commercial†	Bronze
Bronze, Leaded	Bronze
Bronze, Manganese†	Bronze
Bronze, Phosphor†	Bronze
Bronze, Roman†	Bronze
Bronze, Silicon†	Everdur
Bronze, Tobin†	Bronze
Cast Iron*	Cast Iron
Cast Iron, Gray*	Cast Iron
Cast Iron, Malleable*	Cast Iron
Cast Steel*	Steel (1-c)
Chrome-Moly Steel*†	Steel (1-c)
Chrome-Nickel Steel*†	Steel Stainless
Copper*†	Copper
Copper, Beryllium	Silver Brazing
Copper, Deoxidized*†	Copper
Copper, Tough Pitch†	Copper
Cupro-Nickel*†	Bronze
Die Castings	White Metal
Dowmetal	Magnesium
Everdur†	Everdur
Herculoy†	Everdur
Inconel*†	Inconel
K Monel*†	Monel
Lead	Lead

*Also Braze-Welding   †Also Silver Brazing

Table 2-1 Oxyacetylene Weldable Metals (Cont.)

Metal	Jiffy Guide
Magnesium	Magnesium
Mazlo	Magnesium
Monel*†	Monel
Muntz Metal†	Bronze
Nickel*†	Nickel
Nickel Silver†	Bronze
Steel, Galvanized	Braze-Welding
Steel, Low Carbon*†	Steel (l-c)
Steel, High Carbon*†	Steel (h-c)
Steel Pipe*†	Steel (l-c)
Steel Sheet	Steel (h-c)
Steel, Spring*†	Steel (l-c)
Steel, Stainless*†	Steel, Stainless
Steel, Tool*	Silver Brazing
White Metal	White Metal
Wrought Iron*†	Wrought Iron

*Also Braze-Welding   †Also Silver Brazing

Table 2-2 Don't Oxyacetylene Weld

Aluminum 17S	Monel Clad
Aluminum 24S	Nickel Clad
Aluminum Bronze	Phosphor Bronze
Aluminum Clad	Stainless Clad
Beryllium Copper	Super Nickel Clad
Chrome-Manganese Steel	Titanium
Inconel Clad	Tungsten-Vanadium Steels

Guides. Those metals not suited to oxyacetylene welding are listed in Table 2-2.

Dissimilar metals may be joined by fusion welding, braze-welding or silver brazing. In some instances more than one process may be used for the joining of dissimilar metals by the oxyacetylene process. Table 3-1 illustrates in chart form the oxyacetylene process which may be used to join one metal to another.

31

Aluminum (Heat-treatable)   *Melting Point—1218 F*

Procedure	*What to do*
Preweld Cleaning	Remove grease, oil and dirt with chemical solvent. If edges are heavily oxidized, pickle in acid, usually wire brushing with hot water is sufficient.
Joint	Same as for steel, beveling over 1/4 in. thickness. Also desirable to notch edges above 1/8 in. Tack weld at 2 in. intervals on 1/16 in. thickness up to 10 in. intervals on 1/2 in. stock.
Position	Downhand preferred, weldable all positions.
Jigs and Fixtures	Desirable for alignment during tacking. It is imperative that parts be free to expand and contract since these alloys are more sensitive to cracking adjacent to weld when cooling. If cracks occur, prevent contraction from placing stresses on the weld.
Preheat	Preheat sheets over 3/8 in. to 650-750 F and hold heat during welding. Do not heat over 750 F.
Rod	Aluminum rod containing 5% silicon.
Flux	Good proprietary powdered fluxes are available. Mix flux with water to form a paste and paint rod and seams on both sides. Stir flux frequently to keep thoroughly mixed.
Flame	Excess acetylene. Use tip same size or one size larger than for the same thickness of steel.
Welding Technique	Puddle continually. Do not over heat as aluminum shows no color indication of reaching melting temperature. Get complete penetration. Use forehand welding. Complete weld in one pass.
Cleaning	Be sure to remove slag and adhering flux by washing in 5% sulphuric acid at 150 F for 10 minutes. Rinse with clear water. Grind and polish weld if invisible joint is desired.
Post Heating	Should be heat treated by heating to 970 F and quenched if possible to restore original strength. If this cannot be done the heat-affected metal in the weld zone will be weaker than base metal.

JIFFY WELDING GUIDES

Aluminum (2S)   *Melting Point—1218 F*

Procedure	*What to do*
Preweld Cleaning	Remove grease, oil and dirt with chemical solvent. If edges are heavily oxidized, pickle in acid, usually though, wire brushing with no water is sufficient.
Joint Prepara- tion	Same as for steel beveling for over ¼ in. thickness. Also desirable to notch edges on thickness above ⅛ in. Tack weld at 2 in. intervals on ⅛ in. thickness up to 10 in. intervals on ½ in. stock.
Position	Downhand preferred though weldable in all positions.
Jigs	Desirable for alignment and to prevent collapse during welding.
Preheat	Preheat sheets over ⅜ in. to 650-750 F and hold at this temperature during welding. Do not heat over 750 F.
Rod	Aluminum rod same purity as base metal.
Flux	A good flux is important and proprietary powdered fluxes are available. Mix flux with water to form a paste and paint rod and seams on both sides. Stir flux paste frequently to keep thoroughly mixed.
Flame	Excess acetylene. Use tip same size or one size larger than for the same thickness of steel.
Welding Technique	Puddle continually the same as for steel. Do not over heat as aluminum shows no color indication of reaching melting temperature. Be sure to get complete penetration. Use forehand welding.
Cleaning	Be sure to remove slag and adhering flux by washing in hot water in 5% sulphuric acid at 150 F for 10 minutes. Rinse with clear water. Grind and polish weld, if invisible joint is desired.
Postheat	Not necessary to postheat.

33

B r a s s (Copper-Zinc)   *Melting Point*—1652 F

Procedure	*What to do*
Preweld Cleaning	Remove oil, grease and dirt by solvents.
Joint Prepara- tion	Same type of edge preparation as for same thickness of steel. Grooved steel or asbestos back-up bars should be used because of high fluidity of brass.
Position	Downhand or vertical. Molds are sometimes used to assist in vertical welds.
Jigs and Fixtures	Desirable but all points of contact between metal and jig should be insulated by asbestos to prevent heat loss. Loosen jig as metal begins to cool.
Preheat	Preheat of 420-480 F is desirable. Efforts should be made to prevent chilling of metal due to rapid dissipation of heat.
Rod	A rod similar in composition to metal being welded or a good bronze rod.
Flux	A good flux is important and proprietary powdered fluxes are available. Mix flux with water to form a paste and paint rod and seams on both sides. Stir flux paste frequently to keep thoroughly mixed.
Flame	Oxidizing. Use flame sufficiently oxidizing to prevent fuming of zinc. Use tip one or two sizes larger than for same thickness of steel.
Welding Technique	Either forehand or backhand technique with forehand preferred. Puddle continually and do not overheat. Overheating will burn out zinc to ruin the base metal. Be sure puddle is molten before attempting to add filler metal.
Cleaning	Remove slag by chipping and brushing.
Postheat	Not necessary to postheat.

JIFFY WELDING GUIDES

Braze-Welding   *Melting Point (of rod)*—1625 F

Procedure	*What to do*
Preweld Cleaning	Clean thoroughly! Remove oil, grease or paint by chemical action or flame-cleaning. Also wire brush, sand blast or grind surfaces to insure cleanliness.
Joint	Butt joints preferred. Bevel edges by chipping, grinding or flame cutting. If flame-cut, clean bevel by sand blasting or grinding.
Position	Downhand best, other positions extremely difficult.
Jigs	Desirable to hold broken pieces in alignment.
Preheat	No preheat required though preheats up to 400 F may be advantageous.
Rod	Good bronze rod—naval brass, manganese bronze or high zinc brass.
Flux	Use a good flux prepared for braze-welding. A "highly oxidizing" brazing flux designed especially for removing carbon or graphite from the surface of the cast iron. Usually manganese dioxide is in such a flux as an oxidizing agent. Dip rod in flux often.
Flame	Neutral or slightly reducing. Use tip one size larger than recommended for same thickness mild steel.
Welding Technique	Backhand technique. Do not melt base metal, heat to dark red (700 F) only. Keep rod fluxed and in envelope of flame, as it melts apply to both sides of break to "tin." If base metal is too hot, filler metal will bubble; if too cold, filler metal will ball up and not tin. Base metal will not tin unless clean. Unless the tinning is satisfactory it will be impossible to make a satisfactory braze-weld.
Cleaning	Remove slag by chipping and brushing.
Postheat	Cover with asbestos to keep from chilling.

35

B r o n z e (Copper-Tin) *Melting Point*—1625 F

Procedure	*What to do*
Preweld Cleaning	Remove oil, grease and dirt by solvents.
Joint Prepara-tion	Same type of edge preparation as for same thickness of steel. Grooved steel or asbestos back-up bars should be used because of high fluidity of bronze.
Position	Downhand or vertical. Molds are sometimes used to assist in vertical welds.
Jigs and Fixtures	Desirable but all points of contact between metal and jig should be insulated by asbestos to prevent heat loss. Loosen jig as metal begins to cool.
Preheat	Preheat of 420-480 F is desirable. Efforts should should be made to prevent chilling of metal due to rapid dissipation of heat.
Rod	Similar to base metal if color match is desired or a good bronze welding rod.
Flux	A good flux is important and proprietary powdered fluxes are available. Apply to weld puddle by dipping end of heated rod into flux.
Flame Adjust-ment	Oxidizing. Flame should be sufficiently oxidizing to retard melting until red heat is reached but not so much so as to form an oxide coating on the weld puddle. Tip one or two sizes larger than same thickness of steel.
Welding Technique	Control puddle carefully by manipulation of rod and flame. Be sure to melt out tack welds. Be sure puddle is molten before attempting to add filler metal.
Cleaning	Remove slag by chipping and brushing.
Postheat	Not necessary to postheat.

JIFFY WELDING GUIDES

Cast Iron (Fusion welding)   *Melting Point—2300 F*

Procedure	*What to do*
Preweld Cleaning	Clean thoroughly! Remove oil, grease or paint by chemical action or flame-cleaning. Also wire brush, sand blast or grind surfaces to insure cleanliness.
Joint	Butt joints preferred. Bevel edges by chipping, grinding or flame cutting. If flame-cut, clean bevel by sand blasting or grinding.
Position	Downhand best, other positions extremely difficult.
Jigs	Jigs are often desirable to hold broken pieces in alignment. Carbon paste and blocks are also helpful.
Preheat	Preheat entire casting to 1000-1200 F and maintain at this heat during the entire welding operation.
Rod	Cast iron welding rod, preferably one of silicon content.
Flux	Never try to weld cast iron without using a flux! Good proprietary fluxes are available—ordinarily these fluxes have borax bases containing alkali salts to aid in fluxing the slag which forms on cast iron.
Flame	Neutral. Use a tip one size larger than recommended for same thickness mild steel.
Welding Technique	Puddle continually. Never dip a cold welding rod into the puddle. Once a weld has been started continue until finished. Do not rework weld without adding weld metal. Forehand welding preferred.
Cleaning	Remove slag by chipping and wire brushing.
Post Heating	Postheat to 1100-1500 F. Small pieces may then be allowed to cool in a pack annealing pit. Large pieces should be covered with asbestos in the preheating furnace until the fire dies and the piece has become cold.

37

Copper (Deoxidized)   *Melting Point—*1980 F

Procedure	**What to do**
Preweld Cleaning	Remove oil, grease and dirt by solvents.
Joint Preparation	Same type of edge preparation as for same thickness of steel. Grooved steel or asbestos back-up bars should be used because of high fluidity of copper.
Position	Downhand or vertical. Molds are sometimes used to assist in vertical welds.
Jigs and Fixtures	Desirable but all points of contact between metal and jig should be insulated by asbestos to prevent heat loss. Loosen jig as metal begins to cool.
Preheat	Preheat to 420-480 F desirable. Efforts should be made to prevent chilling of metal due to rapid dissipation of heat.
Welding Rod	Deoxidized copper rod where color match or corrosion resistance is desired, otherwise good bronze rod.
Flux	Not necessary, but bronze flux sometimes makes job easier.
Flame	Neutral. Use tip one to two sizes larger than for same thickness of steel.
Welding Technique	Control puddle carefully by manipulation of rod and flame. Be sure to melt out tack welds. Be sure puddle is molten before attempting to add filler metal. On heavy gauges start weld away from edge and weld to one edge, then go back to starting point and weld to other edge.
Cleaning	Wire brushing to clean weld.
Post Heating	Peen welds thoroughly to reduce grain size then heat to 1050-1150 F to restore ductility and improve strength and hardness.

38

JIFFY WELDING GUIDES

E v e r d u r (Copper-Silicon)   *Melting Point*—1866 F

Procedure	*What to do*
Preweld Cleaning	Remove oil, grease and dirt by solvents.
Joint	Same type of edge preparation as for same thickness of steel. Grooved steel or asbestos back-up bars should be used because of high fluidity of copper.
Position	Downhand or vertical. Molds are sometimes used to assist in vertical welds.
Jigs and Fixtures	Desirable, but all points of contact between metal and jig should be insulated by asbestos to prevent heat loss. Loosen jig as metal begins to cool.
Preheat	Preheat of 420-480 F is desirable. Efforts should be made to prevent chilling of metal due to rapid dissipation of heat.
Rod	Use a copper-silicon rod of composition similar to base metal.
Flux	A good flux is important and proprietary powdered fluxes are available. Apply to weld puddle by dipping end of heated rod into flux.
Flame	Neutral or slightly oxidizing. Use tip one size larger than recommended for same thickness of steel. A small concentrated flame is desired.
Welding Technique	Control puddle carefully by manipulation of rod and flame. Make weld as quickly as possible with flame concentrated on small weld puddle. Weld up to ⅜" thickness in one pass. Start welds on thick gauges away from edge and weld to edge, then finish weld by returning to original starting point and welding to other edge. Use same sequence on two-pass welds.
Cleaning	Remove slag by chipping and brushing.
Post Heating	Not necessary to postheat. Mechanical properties improved by hot peening. Grain structure improved by cold peening and annealing.

39

Hard Facing

Procedure	*What to do*
Preweld Cleaning	Remove grease, oil, dirt and rust by grinding or vigorous wire brushing. Surfaces must be thoroughly cleaned.
Position	Hard facing should be applied in flat position.
Fixtures	Desirable to place parts in flat position.
Preheat	Best to preheat to 300-400 F. Thick parts to be hard faced may require preheats to 800 F.
Hard Facing Rod	There are a variety of hard facing rods to meet different wear conditions. Select hard facing rod to resist particular abrasive conditions and to produce desired hardness.
Flux	Good proprietary fluxes are available and often desirable to insure cleanliness.
Flame	Excess acetylene. Have an excess acetylene feather 2 to 3 times the length of the inner cone. Use medium size tip (drill size 62 to 48).
Hard Facing Technique	Backhand technique preferred. For steel: Direct flame on surface at angle of 30 to 60 deg. with inner cone about 1/8 in. from metal. Heat until thin surface layer begins to melt or "sweat". Place rod in flame and melt on sweated area. Puddle deposit to insure smooth, uniform deposit; melting more rod as required. Make deposit in one pass, if possible, then remove flame slowly to prevent cracks or shrink holes. For cast iron: Use flux and weld on thin first layer; then deposit as on steel. On stainless steel: Same technique as for steel except for Type 321. On Type 321 face with Type 347 then hard face.
Cleaning	Remove slag by chipping or wire brushing.
Postheat	Not always required but all deposits should be cooled slowly. Parts which tend to crack should be furnace heated (while still hot) to 1150 F then cooled slowly in closed furnace.

JIFFY WELDING GUIDES

Inconel   *Melting Point—2540 F*

Procedure	*What to do*
Preweld Cleaning	Remove oil, grease and dirt with solvents to insure carefully cleaned surface.
Joint	Any type joint satisfactory. Prepare as for mild steel with groove angle 75 deg.
Position	Weldable in any position.
Jigs	Desirable on thin sections. Use chill blocks near joints and grooved back up bars.
Preheat	Not necessary to preheat.
Rod	Use Inconel rod.
Flux	Special Inconel flux dissolved in an alcohol-shellac solution. Paint rod and both sides of edges to be welded. Keep flux in non-metallic container.
Flame	Slightly excess acetylene. Use tip one size larger than recommended for same thickness of mild steel. Use soft flame.
Welding Technique	Do not puddle. Remember Inconel is a sluggish metal when melted. Keep rod within outer flame envelope to avoid oxidization. Touch rod to surface of weld pool and allow filler metal to flow smoothly into pool. Weld in one pass and do not reweld. Forehand welding preferred. Inconel is more sluggish than nickel, less sluggish than Monel.
Cleaning	Remove flux by sandblasting or immersion of part in 50% nitric acid bath for 10 minutes, then rinse in clear water.
Postheat	Not necessary to postheat.

41

Lead   *Melting Point— 621 F*

Procedure	*What to do*
Preweld Cleaning	Remove oil, grease and other foreign materials; wire brush and scrape clean to remove oxide.
Joint Prepara- tion	Butt joints preferred. Lap joints weld both sides. Edges may be beveled by melting and scraping out molten base metal to a knife edge.
Position	Downhand preferred though lead may be welded in any position.
Jigs	Often desirable but not necessary.
Preheat	Preheating not required.
Rod	Lead rod or strips of lead.
Flux	No flux required.
Flame	Excess acetylene. Use small tip.
Welding Technique	Direct flame perpendicular on base metal. Heat lead until it begins to melt with inner cone of flame almost touching metal. Keep puddle small adding filler metal a drop at a time across joint. Do not overheat or allow puddle to get too large. Be sure to get complete fusion especially on multiple layers.
Cleaning	No cleaning required.
Postheat	Postheat treatment is not necessary.

M a g n e s i u m   *Melting Point*—1240 F

Procedure	*What to do*
Preweld Cleaning	Remove grease, oil and dirt with chemical solvent. If edges are heavily oxidized, pickle in acid, often wire brushing in hot water is sufficient.
Joint	Same as for steel, beveling for over ¼ in. thickness. Also desirable to notch edges on thickness above ⅛ in. Tack weld at 2 in. intervals on ⅟₁₆ in. thickness up to 10 in. intervals on ½ in. stock. Avoid all types of lap joints.
Position	Downhand preferred, weldable in all positions.
Jigs and Fixtures	Desirable for alignment during tacking operation. It is imperative that the parts be free to expand and contract since these alloys are more sensitive to cracking adjacent to the weld bead when cooling. If cracks occur, steps must be taken to prevent contraction from placing stress on weld.
Preheat	Preheat sheets over ⅜ in. to 650-750 F and hold heat during welding. Do not heat over 750 F.
Rod	Use welding rod same as base metal.
Flux	Good proprietary powdered fluxes are available. Mix flux with water to form a paste and paint rod and edges on both sides. Stir flux frequently to keep thoroughly mixed. Use only minimum amount of flux.
Flame	Neutral. Use tip same size or a size larger than for same thickness of steel.
Welding Technique	Puddle continually. Do not overheat as magnesium shows no color indication of reaching melting temperature. Get complete penetration. Use forehand welding. Complete weld in one pass.
Cleaning	Remove slag and flux by scrubbing with wire brushes and hot water. If all flux cannot be removed by brushing, boil in 0.5% sodium-dichromate solution for 2 hrs. Rinse with clear water.
Postheat	Not necessary to postheat.

M o n e l   *Melting Point—2370 F*

Procedure	*What to do*
Preweld Cleaning	Remove oil, grease and dirt with solvents to insure carefully cleaned surface.
Joint	Any type joint satisfactory. Prepare as for mild steel with groove angle 75 deg.
Position	Weldable in any position.
Jigs	Desirable on thin sections. Use chill blocks near joints and grooved back up bars.
Preheat	Not necessary to preheat.
Rod	Monel welding rod similar composition to base metal.
Flux	Use special flux for Monel. Make thin paste by dissolving flux powder in alcohol or boiling water, paint on rod and both sides of edges to be welded. Keep flux in non-metallic container.
Flame	Slightly excess acetylene. Use tip one size larger than recommended for same thickness of mild steel. Use soft flame.
Welding Technique	Do not puddle. Keep rod within outer flame envelope to avoid oxidization. Touch rod to surface of weld pool and allow filler metal to flow smoothly into pool. Weld in one pass and do not reweld. Forehand welding preferred. Monel metal flows quite freeely when melted.
Cleaning	Remove flux by wire brushing and hot water.
Postheat	Not necessary to postheat.

44

JIFFY WELDING GUIDES

Nickel *Melting Point—2646 F*

Procedure	*What to do*
Preweld Cleaning	Remove oil, grease and dirt with solvents to insure carefully cleaned surface.
Joint	Any type joint satisfactory. Prepare as for mild steel with groove angle 75 deg.
Position	Weldable in any position.
Jigs	Desirable on thin sections. Use chill blocks near joints and grooved back up bars.
Preheat	Not necessary to preheat.
Rod	Nickel welding rod similar to base metal.
Flux	No flux needed.
Flame	Slightly excess acetylene. Use tip one size larger than recommended for same thickness of mild steel. Use soft flame.
Welding Technique	Do not puddle. Remember nickel is a sluggish metal when melted. Keep rod within outer flame envelope to avoid oxidization. Touch rod to surface of weld pool and allow filler metal to flow smoothly into pool. Weld in one pass and do not reweld. Forehand welding preferred.
Cleaning	Wire brush to clean.
Postheat	Not necessary to postheat.

45

Silver Brazing   *Melting Point—1250 F**

Procedure	*What to do*
Preweld Cleaning	Remove all oil, grease and foreign material. Clean joint surfaces by use of emery cloth or by pickling to remove scale or highly polished surfaces.
Joint Preparation	Lap joints preferred though butt joints are possible. Grind or machine joint surfaces so that they may be held parallel and equidistant from each other. Joint spacing should be limited between 0.003 to 0.006 in. Thin brazed joints are the strongest.
Position	Brazing may be done in any position.
Jigs and Fixtures	Desirable to hold parts in proper positions. Also desirable during cooling to keep strain from brazed joint.
Preheat	Preheating not required.
Welding Material	Round or rectangular wire, sheets, washers or powdered silver brazing alloys may be used depending upon type of joint.
Flux	A good flux is important and proprietary fluxes, in paste form, are available. Brush along joint to insure proper coverage.
Flame	Excess acetylene. Size of tip will depend upon thickness of metal at joint. Multiple flame torch sometimes desirable.
Welding Technique	Keep torch in motion. Play flame on base metal to let the heat in the joint melt the brazing alloy. Be sure to heat joint surfaces to a point at which the brazing alloy flows freely before attempting to feed alloy into joint. Do not overheat!
Cleaning	Wire brush if necessary.
Postheat	Post heating not required.

*Melting points vary from 1125 to 1760 F depending on alloy.

Steel, High Carbon (Above 0.35% C)   *M. P.* 2500 F

Procedure	*What to do*
Preweld Cleaning	Remove grease, oil and similar foreign matter by chemical solvents or flame cleaning. Remove rust, scale and dirt by wire brushing, chipping or grinding.
Joint Preparation	Butt joint preferred but any joint may be satisfactorily welded. Prepare joint according to thickness of metal as indicated on page 20. Bevel by chipping, grinding or flame cutting.
Position	May be welded in any position.
Jigs	Not necessary unless desired for alignment purposes.
Preheat	Preheating to 300 500 F, depending upon carbon content, highly desirable.
Rod	Alloy steel welding rods desirable, such as, AWS Specification GB65.
Flux	Flux not necessary.
Flame	Excess acetylene. Same size tip as for same thickness low carbon steel.
Welding Technique	Backhand technique preferred. Avoid overheating which is indicated by bubbles in weld puddle. Work bubbles, if any, out to avoid blowholes. Finish weld quickly. Do not remelt or rework.
Cleaning	Wire brush to remove scale.
Post Heating	Post heating may be required to restore special properties originally obtained by heat treatment of high carbon steel.

47

Steel, Low Carbon (Mild) $M. P.$—2700 F

Procedure	*What to do*
Preweld Cleaning	Remove grease, oil and similar foreign matter by chemical solvents or flame cleaning. Remove rust, scale and dirt by wire brushing, chipping or grinding.
Joint Preparation	Butt joint preferred, but any joint may be satisfactorily welded. Prepare joint according to thickness of metal as indicated on page 20. Bevel by chipping, grinding or flame cutting.
Position	May be welded in any position.
Jigs	Not necessary to use jigs.
Preheat	Not necessary. On heavy sections a 400 to 700 F preheat will reduce welding time.
Rod	Use low carbon steel welding rod meeting AWS Specification GA-60 or GB-60.
Flux	Flux not necessary.
Flame	Neutral. Use tip size recommended for low alloy steels as given on page 15.
Welding Technique	Backhand or forehand technique as desired. Inner cone of torch should never touch weld puddle. Keep rod in puddle, moving it from side to side as melting takes place.
Cleaning	Wire brushing will remove scale.
Postheat	Post heating not required.

JIFFY WELDING GUIDES

Steel, Stainless (18-8)   *Melting Point—2550 F*

Procedure	*What to do*
Preweld Cleaning	Remove oil, grease and dirt with solvents to insure carefully cleaned surface.
Joint	Any type joint satisfactory. Prepare as for mild steel.
Position	Weldable in any position.
Jigs	Desirable on thin sections. Use chill blocks near joint and grooved back-up bars.
Preheat	Preheating not ordinarily necessary. For thicknesses over 1 in. slight preheat may be desirable to remove chill.
Welding Rod	Stainless steel (18-8) welding rod, preferably one alloyed or "stabilized" with columbium, titanium, molybdenum, or tungsten to avoid damaging corrosion resisting properties.
Flux	Good stainless steel flux absolutely necessary. Paint rod with flux as well as top and bottom sides of weld joint.
Flame	Slightly excess acetylene. Use tip one size smaller than recommended for same thickness mild steel.
Welding Technique	Hold flame so that it strikes plate from nearly vertical position with inner cone about $\frac{1}{8}$ in. from puddle. Keep puddle small and do not stir. Keep rod near inner cone and weld with steady, even speed. If spacing between plates starts to close, welding speed is too fast, if it starts to open, speed is too slow.
Cleaning	Remove slag and excess flux chemically or by wire brush. Grind and polish weld.
Postheat	Post heating not required.

White Metal*   *Melting Point*— 800 F†

Procedure	What to do
Preweld Cleaning	Clean thoroughly! Remove oil, grease or paint by chemical action or flame-cleaning. Also wire brush, sand blast or grind surfaces to insure cleanliness. File away chrome or nickel plating.
Joint	Butt joint preferred. Bevel edge by grinding or filing. May be melted out with torch.
Position	Downhand best, other positions extremely difficult.
Jigs and Fixtures	Extremely desirable to jig parts to maintenance of alignment and to prevent sagging. On thin sections support with carbon putty.
Preheat	Warm. Best plan is to place parts to be welded on ⅜ in. or ½ in. steel plate and heat plate to 600 F by applying torch flame to underneath side.
Rod	White metal welding rod. Clean rod of oxide coating before using.
Flux	No flux necessary.
Flame	Considerable excess acetylene. Use No. 1 tip for all thicknesses.
Welding Technique	Backhand or forehand. Hold torch tip ½ in. or ¾ in. away from base metal. Heat base metal until it starts to flow. Turn flame parallel to surface and hold heat with side of flame while heating rod. When rod and base metal are molten touch rod to break and rod will flow into use. Dip rod into puddle to break surface tension and insure fusion.
Cleaning	Wire brush to remove scale.
Postheat	No after weld treatment required.

*Some white metal coatings cannot be welded. †Zinc die casting melts at 725 F; aluminum and magnesium die casting melts at 1080-1195 F.

JIFFY WELDING GUIDES

Wrought Iron   *Melting Point—2750 F*

Procedure	*What to do*
Preweld Cleaning	Clean Thoroughly! Remove oil, grease or paint by chemical action or flame-cleaning. Also wire brush, sand blast or grind surfaces to insure cleanliness.
Joint	Any type joint satisfactory. Prepare as for mild steel.
Position	Any position of welding satisfactory.
Jigs	Not necessary to use jigs.
Preheat	Not necessary but 700 to 900 F preheat will decrease welding time.
Rod	Good low carbon rod—AWS Specification GA60 or GB60.
Flux	Flux is not required.
Flame	Neutral. Same size tip as for similar thickness of mild steel.
Welding Technique	Keep large puddle. Be sure base metal is melted, greasy appearance of melting slag at 2200 F fools some. Do not mix base metal and weld metal excessively. Backhand technique.
Cleaning	Wire brush to clean.
Postheat	Not necessary to postheat.

- --- -

51

Table 3-1

	Wrought Iron	White Metal	Stainless Steel	Nickel	Monel	Manganese Steel	Malleable Iron
Aluminum							
Brass							
Bronze							
Cast Iron							
Copper							
Galvanized Steel							
High-Carbon Steel							
Inconel							
Low-Carbon Steel							
Magnesium							
Malleable Iron							
Manganese Steel							
Monel							
Nickel							
Stainless Steel							
White Metal							
Wrought Iron							

52

JOINING DISSIMILAR METALS

Jiffy Joining Chart

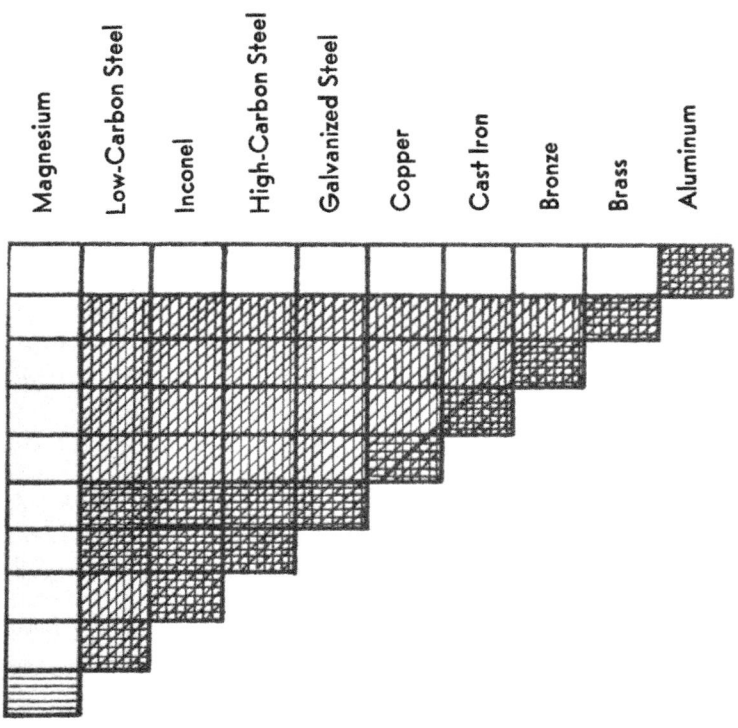

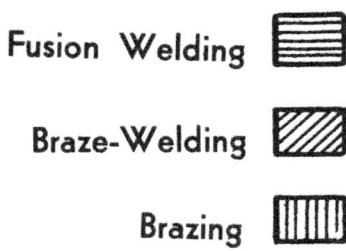

Fusion Welding

Braze-Welding

Brazing

CHAPTER 3

Joining Dissimilar Metals

The problem of welding similar metals is not too difficult when the *Jiffy Welding Charts* are handy. The joining of dissimilar metals, however, is something else. Some metals, for example, magnesium cannot be joined to another metal by welding or an allied process. Nickel, however, may be joined to copper or steel by fusion welding, braze-welding or silver brazing.

Knowing which process to use in the joining of dissimilar metals is a part of the "know-how" which makes the all-around weldor. On the preceding page is a "Jiffy Joining Chart" Table 3-1 which will take many disheartening hours out of the acquiring the know-how of joining of unlike metals. Aided by this *Jiffy Joining Chart* will make it possible to determine quickly the proper means of joining one metal to another by means of the oxyacetylene flame. By checking the key to the markings on the Joining Chart it is easy to determine the process recommended for joining dissimilar metals.

Despite the fact that most metals may be welded by the oxyacetylene process there are some metals for which this means of joining is not recommended. The more commonly used commercial metals of this group are listed in Table 2-2.

CHAPTER 4

How to Avoid Defective Welds

Defective welds are the result of improper welding methods. By becoming acquainted with the various defects which might develop, these faults may be overcome or avoided. Most of the difficulties encountered in welding may be attributed to doing—*some little thing incorrectly.*

If these faults are overcome early, entirely satisfactory welds will be produced. If not, faulty welding habits may be developed which will be quite difficult to discard. The easiest way to determine whether faulty welding habits have been acquired is to examine completed welds and compare them with the examples which follow.

Incomplete Welds. An incomplete weld, commonly referred to as "lack of penetration", is one of the most common welding faults, particularly for the new weldor. This term—*lack of penetration*—is often misunderstood because a weldor may be easily fooled by the appearance of a weld. Lack of penetration usually results from an edge preparation difficulty. The parts to be welded may be spaced too closely or the edge bevel may be incorrect.

A weldor may be deceived by the mere fact that weld metal is dripping from the underside of a joint—this is not proof that a complete weld is being made. In fact, such an observation may be wrong on several counts!

Often the weld penetration obtained is only partially complete.

For example, some areas in a weld joint may be complete —welded from top to bottom—while in other places this is not the case. In these latter spots the filler metal may be separated from the face of the bevel by an oxide film which prevents fusion. Another misleading indication of penetration is caused by seeing the flame go through the joint though the weld metal may not be getting to the bottom of the groove.

Don't be fooled by appearances. Make sure that there is full penetration otherwise the weld won't be satisfactory. *Unless there is complete fusion the entire face of the bevels the weld is incomplete.*

Fig. 4-1. The perfect weld. Note the uniform deposit, the slight reinforcement and full penetration.

Incomplete welds show up readily when tested. Such welds, because of poor penetration, are weak and will not stand as much bending or pulling as the base metal. It is not necessary to test such welds to destruction. Usually, incomplete penetration may be discovered readily by looking at the reverse side of the weld. Such inspections will show whether the parts have been fused to the bottom of the joint.

The ability to get complete penetration may easily be checked by welding a few samples. Weld them using a normal welding procedure, then cut across the weld and observe the appearance of the deposit. Figure 4-2 shows the weld bead appearance and cross-section of a weld lacking penetration.

HOW TO AVOID DEFECTIVE WELDS

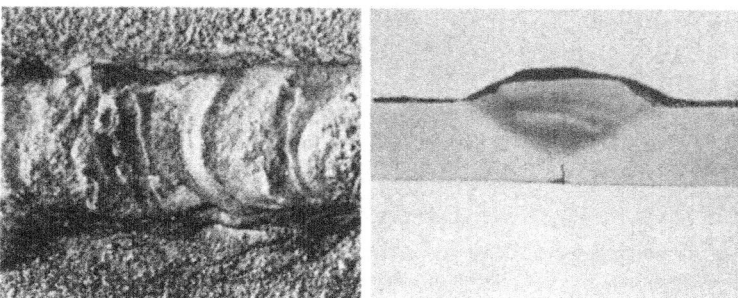

Fig. 4-2. Lack of penetration provides a crack starter leading to weld failure.

Undercutting. An inexperienced weldor is likely to undercut the welds when making a long seam. Though the weld may be perfectly made, undercut occurs when the flame has been improperly manipulated. As a result a portion of the base metal, along side of the weld, is melted away. Sometimes undercut occurs on one side of the weld bead, Fig. 4-3, but generally both sides are undercut, Fig. 4-4. This reduction in thickness of the base metal weakens the joint.

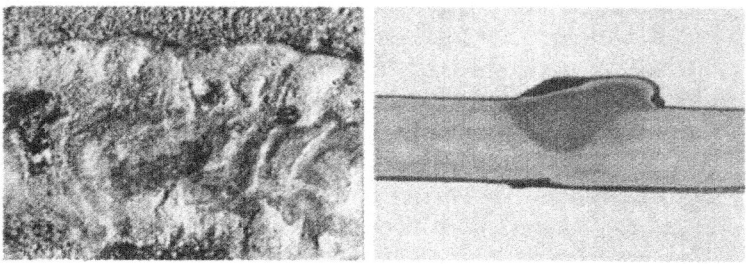

Fig. 4-3. In this weld the base metal has been undercut on one side.

Watch the work carefully to see that neither side of the seam is undercut during the welding operation. There will be no danger of developing this fault provided skill is exercised to obtain an even distribution of heat during welding.

57

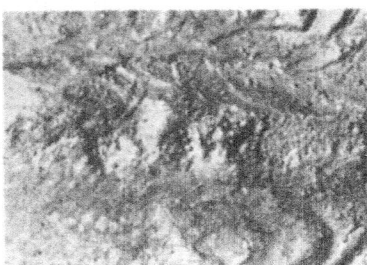 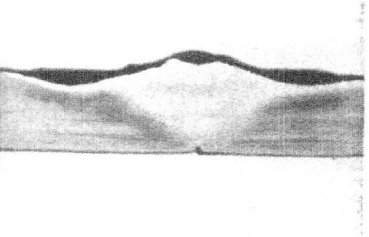

Fig. 4-4. Improper flame direction has resulted in undercut on both sides of the weld bead.

Poor Fusion. Poor fusion is an immediate indication of improper handling of the molten weld metal. It is evidenced by a failure to get clean, sound metal deposited throughout the weld. Poor fusion shows up in one of four forms: laminations, overlap, slag inclusions or oxide inclusions. These difficulties may readily be avoided by following instructions on the proper manipulation of the torch and the welding rod during welding.

Get thoroughly acquainted with the four common causes for poor fusion!

Lamination. Laminations occur when successive layers of weld metal are deposited in the joint without taking care to see that each layer is thoroughly fused to the previous layer. This naturally means that slag and oxides are not floated out of the weld. There may be good, sound weld metal in the joint, but if part of it is made up of layers or drips of weld metal which has not been fused properly, it will have little strength. This fault may be determined readily by subjecting the weld to any type of destructive test. It will show up quite plainly in the fracture.

Overlap. Overlap results when the molten metal or puddle is allowed to flow ahead along the weld joint before the base metal has been thoroughly melted. Unless the base metal is melted it is impossible to obtain fusion. Here again, though the weld may look good, the joint will be of

58

low quality because of the failure to obtain a satisfactory bond.

Slag Inclusions. Slag inclusions are the result of welding on dirty metal or with a dirty filler rod, without taking proper precautions to work the slag to the top of the molten pool. When slag particles are included in the weld metal they weaken the weld just as though the metal has a hole in it.

Oxide Inclusions. Oxide inclusions develop when the weldor fails to work all of the oxides to the top of the molten weld puddle. They produce the same results as slag inclusions.

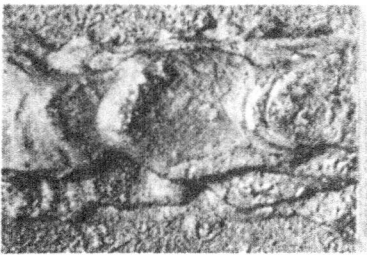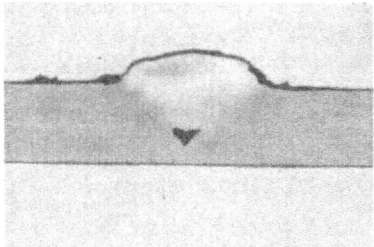

Fig. 4-5. Carelessness leads to defective welds. Incomplete fusion made this weld much weaker.

Inclusions can be found only by examining the interior of the weld metal, Fig. 4-5. If the weld is found to contain dirt, ashes and other foreign material, there are inclusions of some sort.

Oxidized Welds. Oxidized welds are caused by welding with an oxidized flame. Don't confuse an oxidized weld with the occasional oxides which may be present in any weld puddle even when made with a correctly adjusted flame. These oxides need only to be floated or worked to the surface.

Weld metal that has been oxidized looks burnt, Fig. 4-6. It is weak and brittle and must be avoided in important work.

59

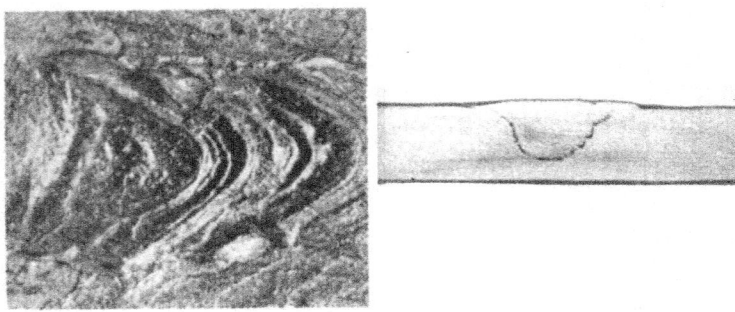

Fig. 4-6. Oxidized welds are burnt looking and brittle.

Oxidized welds are the result of improper flame adjustment, a condition which is not always the fault of the weldor. Sometimes it is impossible to maintain a neutral flame because something is wrong with the regulators or torch. In such cases inspect them carefully to see what is the cause of the difficulties.

Carburized Welds. Carburized welds are caused by using a carburizing flame instead of a neutral flame. They are brittle and are quite easily distinguished from a good weld. A carburized weld may generally be identified by small holes and radial lines extending from the center of the weld bead toward the outside, Fig. 4-7.

To avoid carburized welds, adjust the flame so as to make it carburizing and deliberately make a small carburized weld. Note how the weld metal bubbled and

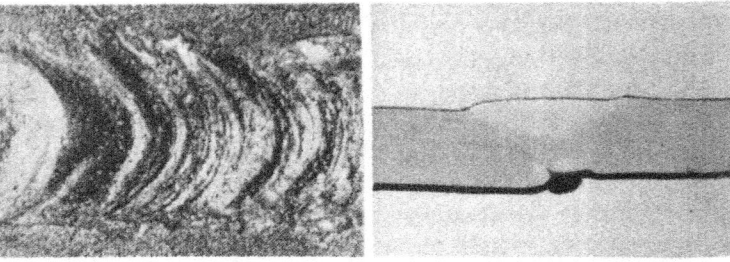

Fig. 4-7. Carburized welds have pin holes and cracks.

60

HOW TO AVOID DEFECTIVE WELDS

sparkled. By recognizing this indication of a carburizing flame it will make it easy to correct the flame to obtain the proper neutral flame for welding.

Hard Spots. Cast iron welding is often made difficult by the occurance of hard spots in the weld. Usually hard spots are not visible on the surface of the weld bead though occasionally there is a small silvery gray spot which indicates such a formation. When a cast iron weld is opened, by breaking or sawing, there are hard spots if the interior is flicked with silver spots indicating formations of white cast iron. White cast iron is extremely hard.

Hard spots in cast iron are very undesirable. They can be found by machining—the resultant broken tools is the major reason for their elimination. They are the starting point for progressive cracks. Needless to say, extreme care should be taken to avoid hard spots. Likewise, the weldor should become familiar with the errors causing these defects.

Hard spots are caused by a sudden chilling or cooling of the weld metal. Putting a cold welding rod into the molten puddle will produce hard spots. Hard spots which show up on the surface as silvery areas may be formed by going over the top of the weld, after it has cooled—remelting small areas which become chilled from cold metal underneath. Hardened areas in the weld may likewise be caused by not having the adjacent metal hot enough to prevent chilling.

To avoid hard spots be sure that the parent metal is kept hot and that a large molten puddle is maintained. This will permit the puddle to solidify slowly as the weld advances. It has been found that ripple type of welding is less apt to produce hard spots than a procedure that calls for successive puddles. Slow cooling of the welded piece will also reduce the tendency for hard spots.

Blow Holes. Blow holes are another difficulty common to cast iron welding. They are the small holes which

61

often may be seen plainly on the surface of the weld. When they appear on the surface it is fairly certain that they will also be present in the interior of the weld.

Blow holes are formed by gasifying particles of foreign material present in the parent metal or filler rod. They may also be caused by the welding method, especially if too small a puddle is used. When the parent metal is not heated sufficiently, the flame must be played directly on to the puddle, this, will not only produce bubbles but force them down into the weld metal to cause blow holes.

Blow holes may be avoided by watching the puddle and floating bubbles out as they appear. This can be done by keeping the puddle large and hot so that the metal under the hole becomes liquid and the gas bubble causing holes comes to the top and disappears.

To eliminate blow holes:

Always keep the base metal well heated near the weld.

Have the puddle large and hot.

Slant the flame ahead so that it does not play upon the rear portion of the puddle as it is solidifying.

Watch the Reinforcement. The mere fact that a weld has adequate reinforcement is no assurance that it has strength. A common fault in oxyacetylene welding is that of building up the weld bead too much. A properly made weld is stronger than the base metal, consequently added

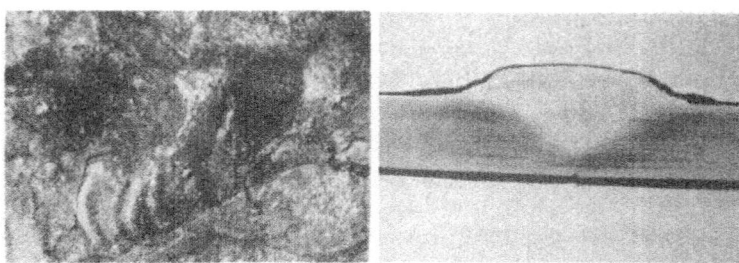

Fig. 4-8. Over reinforcement wastes time and material while contributing nothing to strength.

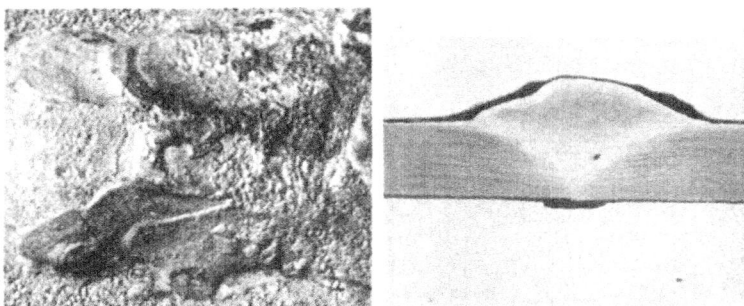

Fig. 4-9. Over reinforcement and incomplete penetration results in a weaker rather than stronger joint.

reinforcement serves no purpose other than wasting time and material.

The weld bead shown in Fig. 4-1 is adequate. There is no reason for depositing a greater amount of weld metal than is indicated in this application. The weld bead pictured in Fig. 4-8 is not pleasing in appearance because it has too much reinforcement. The extra metal contributes nothing to strength though it does increase the weight of the subject being welded. It also increases the problem of finishing the product inasmuch as a great deal more grinding will be necessary to produce a weld joint of satisfactory appearance.

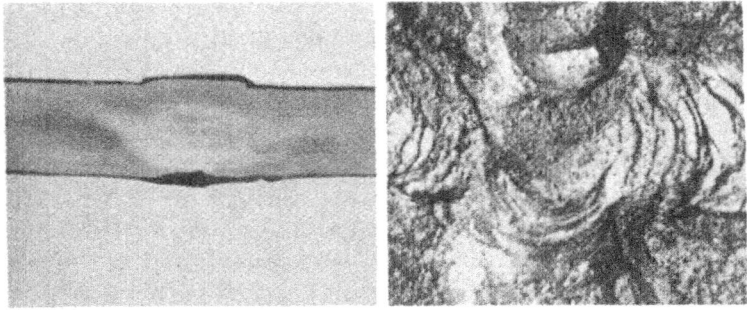

Fig. 4-10. When a weld joins or crosses another it should look like this crossover.

The weld bead pictured in Fig. 4-9 is not only over-reinforced but penetration is incomplete, making the weld extremely unsatisfactory.

Remember the mere piling on of weld metal does not mean that a satisfactory weld is being made.

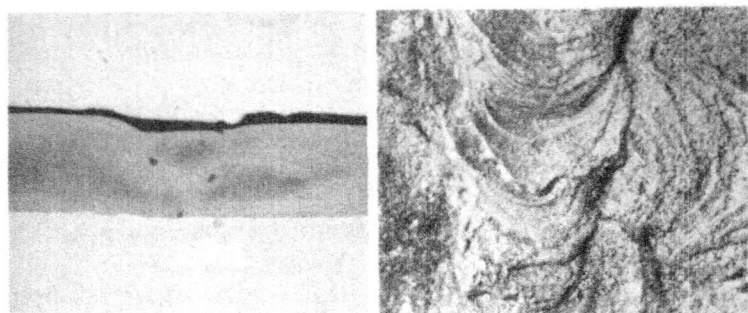

Fig. 4-11. This crossover is unsatisfactory. The weld is rough and filled with blowholes.

Common Defects. Those who are learning to weld should be extremely careful about the type of tack welds they make to join metal pieces. Many of the defects which develop in welding come from a failure to weld through tack welds. Often this defect can be detected only by cutting through completed welds where the tack had been made. A similar difficulty may occur where one welding bead crosses another. Figure 4-10 is an example of a good crossover while Fig. 4-11 pictures the results of a poorly done welding job.

A more obvious welding defect is the failure to maintain a full size weld bead to the end of a seam. This careless act will result in a crater or depression which may easily cause cracking or the complete failure of the weld, Fig. 4-12. Such a fault can be overcome by practice in manipulating the pool of molten metal at the end of the weld.

Crater formation is not necessarily confined to the end of the weld. It may develop any place along the bead

HOW TO AVOID DEFECTIVE WELDS

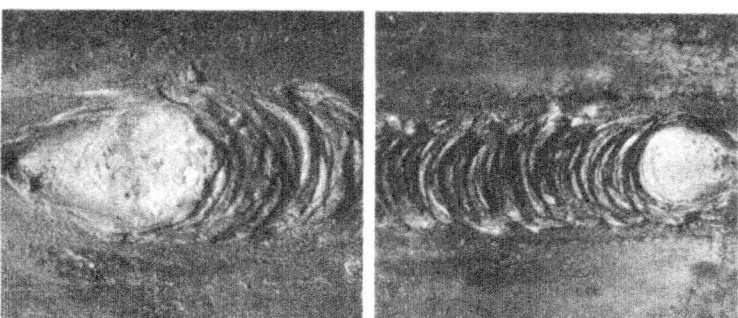

Fig. 4-12.   Craters at the end of a weld bead may mean trouble.

where the welding is stopped long enough to allow the metal to solidify before welding is restarted. Skill in the handling of a welding torch, however, makes it possible to start and stop welds without leaving evidence of any interruption.

So far as the welding faults mentioned here are concerned, they must all be avoided if a perfect weld is to be secured. It is for this reason that the reader should frequently consult this discussion so as to guarantee thorough acquaintance with the faults which must be avoided in actual welding practice.

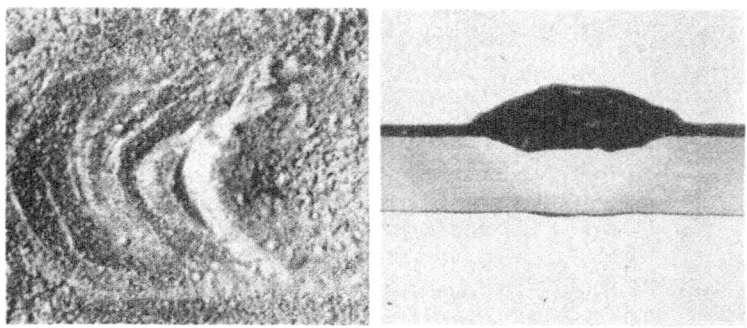

Fig. 4-13.   Skillful torch handling will avoid shallow spots.

Honesty. It is most important that anyone doing welding be absolutely honest with himself and his customers. It is very easy with a torch to gloss over defects, to include particles of slag in the weld, or to weld over the surface and not penetrate to the full depth of the metal. Such practices mean but one thing—a weak, poor weld which will probably break in service and it may cause accident and even death. Any weld can be made stronger than the metal itself by building up the weld thicker than the base plates and by carefully and skillfully handling the torch and the welding rod.

A weldor who is content to gloss over the surface and who will not take the time and trouble to use his skills to the best of his abilities to attain sound, honest welds had better stop right at the beginning and take up some other trade. Welding requires fundamental honesty and conscientious workmanship.

CHAPTER 5

Flame Cutting Steel and Cast Iron

All ferrous metal may be flame cut though some flame cut more readily than others. The technique of cutting is varied with the ferrous material being cut. Generally, the lower the carbon content, the more easily the ferrous material is flame cut. For this reason, low carbon steel is the most easily cut and cast iron the most difficult.

The difference in techniques required to flame cut ferrous metals makes it desirable to break the process down into five distinct types of cutting. Before discussing these various types however, it might be well to review some of the fundamentals of the process.

Be Safe. One of the first steps in a cutting operation is to determine that it is to be a safe operation. Be sure that the material to be cut is placed so it will not fall on your feet upon being severed. Also take care to see that it will not fall on other persons in the vicinity or on the oxygen or acetylene hose of the cutting equipment. Care should also be exercised to protect yourself; others, the hose, and flamable material in the vicinity from the hot slag and sparks which will result from the cutting operation. Particular care must be exercised to see that the hose will not be damaged since the high pressure gases may cause an extremely dangerous condition, should a leak develop.

How to Flame Cut. Having taken the necessary safety precautions, you are now ready to begin cutting. The line of cutting must be established and carefully marked on

67

the material to be cut. A sharp piece of soap stone is preferable for marking the line of cut. Chalk is not desirable inasmuch as the flame will immediately burn it off the metal for some distance. On work where extreme accuracy is required, it is also desirable to mark the line with center punch marks at frequent intervals.

Cutting Tip Selection. Too much emphasis cannot be placed upon the importance of the proper size tip for efficient and economical cutting. All manufacturers of cutting torches have experimented extensively with their torches to determine the tip size best suited for different thicknesses of metal. These experiments, likewise, have determined the recommended acetylene and oxygen pressures for the individual tip size. This information is furnished in chart form. Cutting tip sizes for various thicknesses of steel are given in Table 5-1.

Table 5-1—Cutting Tip Recommendations
(mild steel not preheated)

Thickness of metal, in.	Oxygen Jet Drill Size (center hole)	Oxygen pressure, psig (lb per sq in. gauge)	Acetylene pressure, psig (lb per sq in. gauge)
up to 1/8	60	10	3
1/4	60	10-15	3
3/8	55	15-25	3
1/2	55	20-30	4
3/4	55	20-30	4
1	53	25-35	4
1 1/2	53	30-40	4
2	49	40-50	5
3	49	45-55	5
4	49	50-60	5
5	45	50-70	5-6
6	45	60-80	5-6

The operator should develop a technique which will enable him to get satisfactory cutting speed with the

FLAME CUTTING

Fig. 5-1. When the metal starts to melt turn on the cutting oxygen.

recommended tip size and pressures. These recommendations are usually based on the cutting of comparatively clean metal. When cutting metal which is rusty, scaly or has been painted, it is generally suggested that the next larger size recommended tip be used. Often it will be found more economical to clean the metal along the line of cut so as to expose clean metal.

Starting the Cutting. After assembling the cutting torch, selecting the tip and laying out the cut, you are ready to begin the flame cutting operation. Light the torch and adjust the preheat flames to neutral (about $\frac{3}{16}$ in. long) just as though welding. The preheat flames are now directed on the spot where the cut is to begin until a small spot of metal has been heated to a nearly white heat. Now, withdraw the torch tip so that the tip of the preheat flames are about $\frac{1}{8}$ in. from the metal surface and open the cutting oxygen valve gradually. The process of oxidation will begin immediately so that a small nick (Fig. 5-1) is burned in the metal beneath the torch tip. The heat generated by the oxidation of the metal, aided by the pre-heat flames raises the temperature of the adjacent metal sufficiently to make flame cutting possible. This is done by holding the cutting oxygen valve open while the tip is gradually moved along the line of cut, until the cutting process has been completed.

69

Flame Cutting Variables. There are really ten important variables that will affect the quality of oxyacetylene cutting. They are:

1. Size of tip oxygen jet.
2. Oxygen speed.
3. Thickness of metal.
4. Cutting speed.
5. Purity of oxygen.
6. Quality of the metal.
7. Intensity of preheat.
8. Angle of cutting oxygen stream to metal.
9. Smoothness of cutting oxygen orifice.
10. Cleanliness of tip end.

Flame Cutting Faults. Some of the more common difficulties of flame cutting may be identified by the appearance of the flame cut surface.

Always, follow the recommendations as to the proper tip size and cutting oxygen pressure to be employed for cutting various thicknesses of metal.

The rate of cutting is another factor which must be critically considered in the cutting operation. Remember, it will depend upon the selection of the preheat flame, the correct tip size and adjustment of the cutting oxygen jet. Oxygen must be supplied in the right amounts to properly combine with the metal to be consumed.

Restart Gouges. The beginner at flame cutting often has trouble continuing a steady cut. If the torch is moved too rapidly, the metal does not have a chance to become properly heated. Consequently, the metal will not combine readily with the cutting oxygen and cutting stops. To restart the cut, the metal must again be heated with the preheating flame. When heated the cut may be restarted but unless it is done with care there will be gouges as shown in Fig. 5-2.

Unsteadiness. Another fault of the beginner is the rough, wavy, irregular cut caused by the unsteadiness of

70

operator. A flame cut by an unsteady operator will look similar to that shown in Fig. 5-3.

Too Much Pressure. All control over the cut is lost when the oxygen pressure is too high as may be seen by Fig. 5-4.

Not Enough Oxygen. When the cutting oxygen pressure is too low, the cut has the appearance of that shown in Fig. 5-5. Note how the upper surface has been melted by the preheat flames because the cutting speed was so slow.

Short Preheat Flames. Preheat flames which are too small result in too slow a cutting speed. The resulting cut pictured in Fig. 5-6 is often badly gouged near the bottom of the cut.

Long Preheat Flames. The preheating flames must be just right; neither too long nor too short. In Fig. 5-7 is shown what happens when a flame cut is made using preheating flames which are too long. The cut is quite irregular, the upper surface is melted and there is an excessive amount of slag on the lower surface.

Too Fast. If the cutting speed is too fast the cut will be irregular and the cut lines curve back (Fig. 5-8) rather than stand vertical.

Too Slow. When the torch is moved too slowly, the oxygen has time to combine with the metal on both sides of the kerf as well as the metal directly ahead. As a result the kerf will become much wider than necessary and there will be a considerable waste of welding gases. Figure 5-9 pictures the uneven results of too slow a cutting speed.

Correct Cut. The correct flame cut is easily recognized. The correct application of the cutting flame produces a cut (Fig. 5-10) with square edges and the kerf has parallel sides. An examination of the face shows that the cut lines are vertical uniform and not deep, (Fig. 5-11). In fact, the face of the cut is comparatively smooth and the corners are sharp. There is no evidence of surface melting and little or no adhering slag.

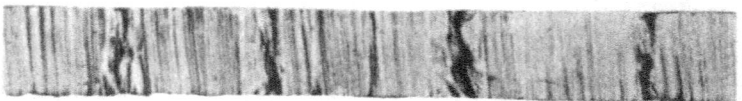

Fig. 5-2. Gouges often result from carelessness in restarting the cut.

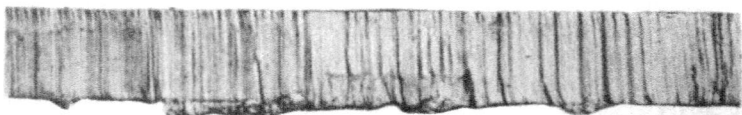

Fig. 5-3. An unsteady operator produces a wavy cut.

Fig. 5-4. Too much pressure causes loss of control.

Fig. 5-5. Cutting oxygen pressure insufficient, so lack of speed causes melting of top edge.

Fig. 5-6. Short preheat flames result in slow cut and groove at bottom edge.

Fig. 5-7. Too long a preheat flame melts the top surface and causes excess slag.

72

Fig. 5-8. Cut lines curve back and edge is uneven when cutting speed is too high.

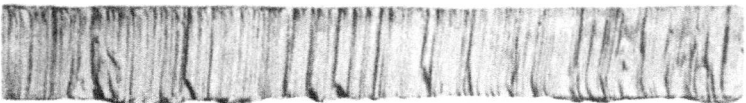

Fig. 5-9. An uneven cut results when cutting speed is too slow.

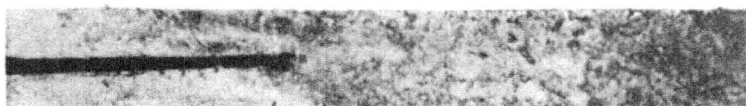

Fig. 5-10. Top view of a good cut. Edges are smooth and parallel.

Fig. 5-11. A correctly made cut has square edges and shallow, vertical cut lines.

Straight Line Cutting. The simplest cut is the straight line cut. When the line of cut has been determined to be a straight line it is desirable to use a guide or straight edge to insure a straight smooth cut. In Fig. 5-12 the operator is using a piece of angle as a guide and steady rest while flame cutting the legs of a number of 8 in. channels.

Bevel Flame Cutting. Fig. 5-13 indicates better than words can describe what happens when insufficient pre-heat is used for bevel-cutting steel plate.

73

Fig. 5-12. An angle serves as a steady rest and guide for straight line cutting.

The reason that more pre-heat is required for bevel-cutting than for vertical-cutting operations is that the torch is at an angle to the plate surface. Because of this, the heat tends to "bounce off" and does not penetrate to the same extent as when the pre-heat flames strike the plate at right angles. Obviously, the flatter the angle of cut, the greater amount of pre-heat required.

There are a number of ways to obtain more heat. For normal beveling up to angles of 45 deg., sufficient pre-heat can be obtained with standard tips by slightly "forcing" the preheat flames as the angle of bevel increases. For very shallow angles it is sometimes necessary to provide supplementary pre-heat, either with a second torch, or by first going over the line of cut with the cutting torch. A size larger tip or slightly slower cutting speeds also are of help sometimes. When selecting the correct size tip, the actual depth of the cut should be considered rather than the thickness of the plate.

74

FLAME CUTTING

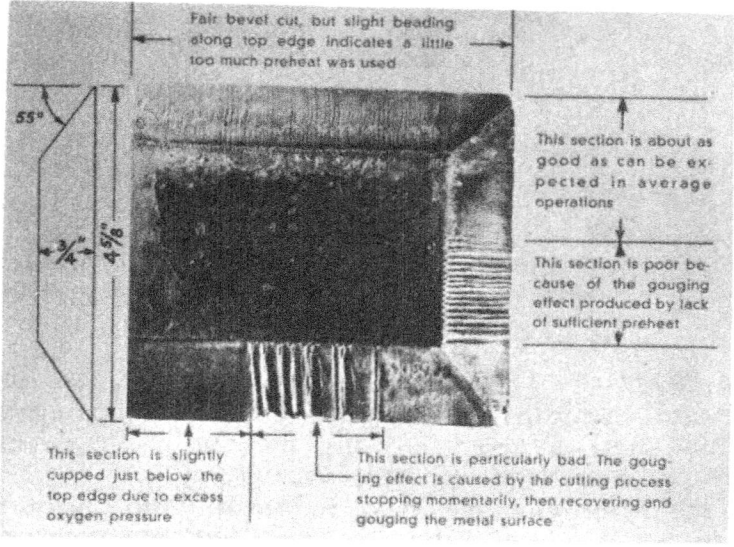

Fair bevel cut, but slight beading along top edge indicates a little too much preheat was used

55°

3/4"

4 5/8"

This section is about as good as can be expected in average operations

This section is poor because of the gouging effect produced by lack of sufficient preheat

This section is slightly cupped just below the top edge due to excess oxygen pressure

This section is particularly bad. The gouging effect is caused by the cutting process stopping momentarily, then recovering and gouging the metal surface

Fig. 5-13. All kinds of errors are made in bevel cutting.

A study of the various cut surfaces in Fig. 5-13 should aid the operator in obtaining maximum results in bevel-cutting operations.

Circle Cutting. For the cutting of circles and rings it is desirable to use a radius arm. There are a number of radius arm attachments on the market for use with the flame cutting torch (Fig. 5-14) or the ingenious worker may make his own.

Flame Cutting Selectors. The International Acetylene Association has developed five types of flame cutting for ferrous metals. It has been determined that as flame cutting shifts from the readily cutable low carbon steels to those of lower cutability, difficulties in the way of making a smooth cut and/or progressing the cut are

Fig. 5-14. A radius arm assists in the flame cutting of curves or circles.

encountered. This gives rise to the need of grading the cutable metals by grouping those of high cutability apart from those of lower cutability and the observance of specific methods of cutting for each group.

At the present time, there are primarily three distinct groups, if viewed from the standpoint of cutability. However, five groups are employed in order to record clearly the various methods of cutting now in use. These are shown on the following pages as Types 1, 2, 3, 4 and 5, wherein the cutable steels are arranged with respect to the methods of technique of flame cutting by which they are severed and which are briefly described.

FLAME CUTTING

Type 1　Flame Cutting
(Conventional Flame Cutting)

Use for Flame Cutting

Low Carbon Steel
Wrought Iron
Rolled or Cast Steels
Firebox
Structural
Steel Forgings*
Cast Steel*
Medium Carbon
Steel*
High Carbon Steel*
Spring Steel*
Tool Steel*
Copper Bearing
Steel
Chromium Steel
Chrome Vanadium
Low Carbon
High Carbon
Tungsten (Low)
Vanadium Steel
Aluminum Steel

Molybdenum Steels
Chrome-Moly*
Nickel Moly
(Low Carbon)
Nickel Moly
(High Carbon)
Manganese Steel
Structural*
Hadfield*
Silico-Manganese*
Silicon Steel
Structural*
Electrical*
Nickel Steel
High Carbon Nickel*
Nickel Chromium
Low Carbon
Medium Carbon*
Chromium Alloy*
Nickel Alloy*

Cutting Technique

Uniform forward movement.
General direction of cut.

The technique of Type 1 cutting begins with a neutral flame preheating of a small area at the start of the cut, until it attains a kindling temperature with oxygen (be-

---

*Heat treatment after cutting desirable to improve cut edge and should be done before piece cools from cutting.

77

tween 1,400 and 1,600 F). When the kindling temperature is attained, combustion of the base metal with the cutting oxygen can proceed. (While theoretically, preheating should not be necessary beyond that of starting the cut, in practice it is maintained throughout the cutting operation; an expedient which insures continuous cutting and high cutting speeds.)

The operation of a cutting torch is merely that of moving it at a practical speed in a line in the direction desired, see above illustration. Control of the cut is possible because of the high velocity of the cutting oxygen stream which operates to confine the cutting oxygen to a very narrow path and, in consequence, its cutting or combustion action is likewise confined.

Remarks

In practice the metal removed from the cut or kerf, is not entirely consumed by the cutting oxygen. A part of it is blown from the kerf in the form of molten metal, the amount depending upon the force of the cutting oxygen stream. This will vary with the depth of the cut as indicated by the manual oxygen pressure tables. The blowing-out of unoxidized metal is responsible to a considerable degree for the fact that in certain cases less oxygen is required to produce a cut than that theoretically needed to oxidize the quantity of metal removed from the kerf. This condition does not apply to thin sheets because with the high velocity of the oxygen and the short path of cut, the opportunity of oxidation and time of contact is too short to permit all of the oxygen to react effectually with the steel.

FLAME CUTTING

Type 2　Flame Cutting
(Flame lancing)

Use for Flame Cutting
Cast Iron
White Cast Iron†
Grey Cast Iron†
Malleable Cast Iron†
Semi-Steel†
Duriron†

Cutting Technique

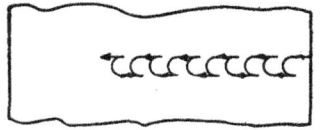

Movement of cutting torch tip for cutting thin cast iron.

Type 2 cutting is known commercially as oxygen lance cutting. Its significant operating differences compared with Type 1 is its use of the heat produced by the combustion of iron (or steel) instead of acetylene to preheat and facilitate the cutting operation. This action is produced by replacing the usual cutting tip with a length of iron (or steel) pipe, through which oxygen only flows. Once kindled, the end of the pipe or lance, as it is called, continues to burn as long as oxygen flows through it. The lance is maintained at the cutting temperature because it is advanced into a pool of melted metal and slag which is constantly replaced as the oxygen reacts with fresh-exposed underlying metal. Only a relatively small quantity of oxygen is consumed in burning the lance, the remainder is available for cutting the base metal. Regulation of the cutting oxygen for different kinds of work is

†Safely cut at room temperatures without subsequent treatment. Carburizing pre-heat flames recommended.

79

attained by varying the pressure at the oxygen cylinder regulators. As lance cutting proceeds, the lance is consumed, ultimately requiring replacement by a new lance or length of pipe.

Remarks

Type 2 cutting is an oxyacetylene cutting means which extends the dimensional or depth range of one standard cutting torch operations. It is used primarily for severing thick or heavy masses of metal. Seldom is it used alone, but generally as an aid to Type 1 cutting. Under proper conditions of work set-up for slag flow and a facility for regular lance technique, however, it can progress independently. Per pound of metal removed, Type 2 usually consumes more oxygen than the other types of cutting. This however, is offset by the greatly increased value of the cutting service compared with other methods that may be used.

The surface produced by Type 2 cutting is not as smooth as is produced by the other types, particularly Type 1.

FLAME CUTTING

Type 3   Flame Cutting
(Cast Iron Cutting)

Use for Flame Cutting

Cast Iron
  White Cast Iron†
  Grey Cast Iron†
  Malleable Cast Iron†
  Semi-Steel†
  Duriron†

High Nickel Steel†
High Speed Steel*
High Tungsten Steel†

Cutting Technique

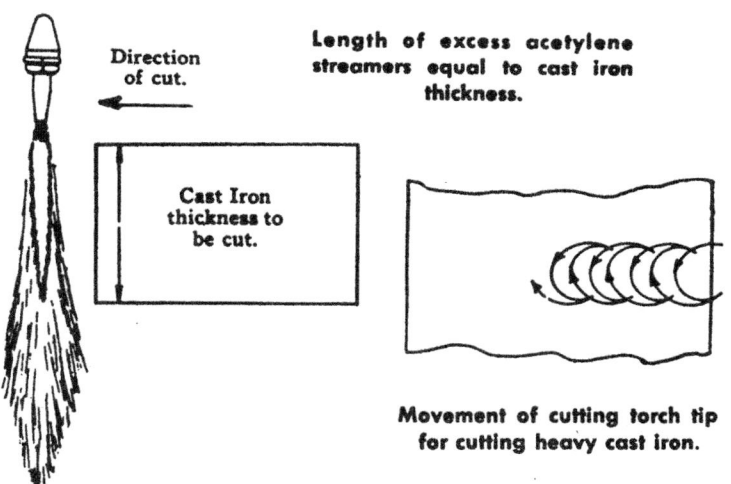

Direction of cut.

Length of excess acetylene streamers equal to cast iron thickness.

Cast Iron thickness to be cut.

Movement of cutting torch tip for cutting heavy cast iron.

This technique was developed for severing the cast irons. It uses considerably more preheat than Type 1 to start the cut and also to maintain it.   A definite flame-oscillation is used during the cutting operation, it is a necessity with the cast iróns, particularly those possessing uncombined or graphitic carbon.

Heat treatment after cutting desirable to improve cut edge and should be done before piece cools from cutting.

†*Safely cut at room' temperature without subsequent treatment.*

81

The preheating flames in Type 3 cutting are highly carburizing. The length of excess acetylene streamer should be approximately equal to the thickness of the cast iron. This flame adjustment provides fuel for preheating beyond the tip of the torch and consequently, aids in maintaining preheat along the path of the cut as the acetylene combines with oxygen from the cutting oxygen stream.

The resulting cut is a "worked cut" for the metal originally in the kerf has, in reality, been "worked out" by the oscillating technique, which, if regularly maintained, causes a constant stream of slag and molten metal to flow through the kerf, preheating the base metal in the kerf. The sidewise motion or oscillation of the flame is in the form of half circles or new-moons at regular and short intervals. A Type 3 cut is rougher, more metal is removed and the oxygen and acetylene consumption is greater than in Type 1 flame cutting.

Remarks

The most essential part of performing Type 3 cutting is the maintenance of a steady and good degree of preheat at the starting point or top of the cut and a steady and uniform slag flow. In the case of the clean homogeneous cast irons, this is not difficult and in consequence, it is possible to make a relatively smooth appearing cut.

FLAME CUTTING

Type 4   Flame Cutting

Use for Flame Cutting
Stainless Steels*
   *Low Carbon Chromium**
   *High Carbon Chromium**
   *Chromium-Nickel**

Cutting Technique

**Short forward movement followed by short reverse movement
repeating steps as indicated by arrows.**

Type 4 cutting may properly be referred to as a counterpart of Type 3 cutting.   The flame is moved continually during the cutting operation, but in. a reciprocating manner in short strokes in the direction of the cut in contrast with the sidewise oscillation used in Type 3.   Its principal use at the present time is for severing stainless steels, which resist oxy-acetylene cutting by the simpler techniques of Types 2 and 3.

In addition to the usual preheating at the point where the cut is started, preheat is also applied along the entire face of the cut until it is brought to a bright red color, whereupon the cutting operation is started and progressed through the part by the forward and backward reciprocating technique described above.

As in Type 3 cutting, the maintenance of a steady stream of liquid slag and molten metal is essential for a successful cut.   Slag erosion probably is as important an agent in making cuts in the stainless steels as any other single function of cutting and determines the degree of uniformity in which cutting in these materials is attainable.

Remarks
Type 4 cutting is rough cutting and calls for subsequent tool machining or grinding if a smooth finish is desired.

*Heat treatment usually is necessary to counteract harmful effects of the heating produced during the cutting operation unless the metal contains a stabilizing element to prevent such heat effects.

Type 5 Flame Cutting
(Flux cutting)

Use for Flame Cutting

Cast Iron†
  White Cast Iron†
  Grey Cast Iron†
  Malleable Cast Iron†
  Semi-Steel†
  Duriron†

High Carbon Nickel Steel
Stainless Steels*
  *Low Carbon Chromium**
  *High Carbon Chromium**
  *Chromium Nickel**
High Tungsten Steels

Cutting Technique

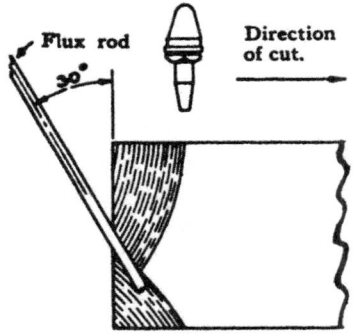

This technique is developed around the use of easily cut metal to aid the cutting of metals of low cutability, just as fluxes are used in certain welding operations to increase the fusibility of refractory oxides. This type of cutting may be done in several ways, but in the main they resort to cover, or cover and intermediate plates of low-carbon steel placed as jackets, or interlaminations between the resistant materials to initiate good cutting conditions and if laminated, to relay these conditions through the stack. The bottom and top jacket plates are sometimes used merely to produce sharp cut edges in

†*Safely cut at room temperatures without subsequent treatment. Carburizing pre-heat flames recommended.*

84

parts otherwise readily cutable, in which case the operation is referred to as stack flame cutting. Metals containing high nickel and chromium can often be cut by Type 5 cutting, while the cutability of all of the lesser resistant materials is greatly improved by its technique.

In another form of Type 5 cutting a low-carbon steel rod or bar is used at the point of cutting to supply additional heat and slag to a cut in a resistant material, thereby supplementing the work of the cutting torch. The bar or rod can be moved at the will of the operator to be available at or adjacent to the critical point of cutting. Castings or resistant materials are often cut quite readily by this means.

Remarks

Any one of the previous four techniques, Type 1, 2, 3 and 4 may be employed in Type 5 cutting, depending upon the specific nature of the resistance. At times it enables materials to be cut by Type 1 cutting that would otherwise require a more tedious technique.

CHAPTER 6

Flame Treating

The oxyacetylene flame is used for many different applications. The most common uses are for welding and flame cutting, but, in addition, there are many other uses—the most important of which are for heating, straightening and cleaning.

Flame Heating. In Chapter 1 there was a brief discussion of the use of the oxyacetylene flame for preheating prior to welding and in Chapter 5 the preheating flames of flame cutting were considered. The oxyacetylene flame is used for other heating purposes, too. In Fig. 6-1 is pictured a flame heating application to simplify the forming of a 1 in. steel plate in the making of a dragline bucket. After the plate has been heated to a red heat it is pulled into place by a chain hoist.

Flame Straightening. Use of the oxyacetylene flame to straighten deformed or bent steel members was pioneered by Joseph Holt of Seattle and he is still the leading authority on the subject. By the Holt method the contraction forces set up in heated metal, depending upon the application of heat, may be used to straighten or bend members as desired.

Ádequate discussion of the subject to give a thorough understanding of the principles of "contraction" straight-

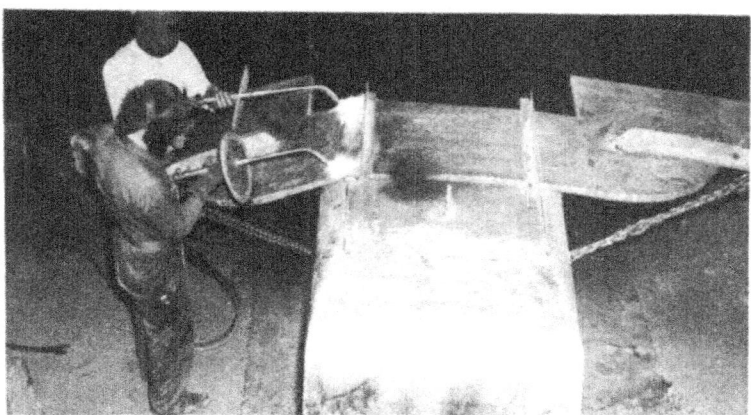

Fig. 6-1.   Flame heating simplifies the forming of 1-in. steel plate.

ening would necessitate writing another book.   The weldor, however, may learn much about the powers of expanding and contracting forces from experimenting. Get various bent pieces of steel and heat them to see if they can be straightened by heat alone.   The results will be amazing.

In Fig. 6-2 a weldor is straightening a bridge tie bar by heating it in a wavy fashion, while the 5 in. diameter shaft

Fig. 6-2.   Heating in a zig-zag fashion straightens a bent bridge member.

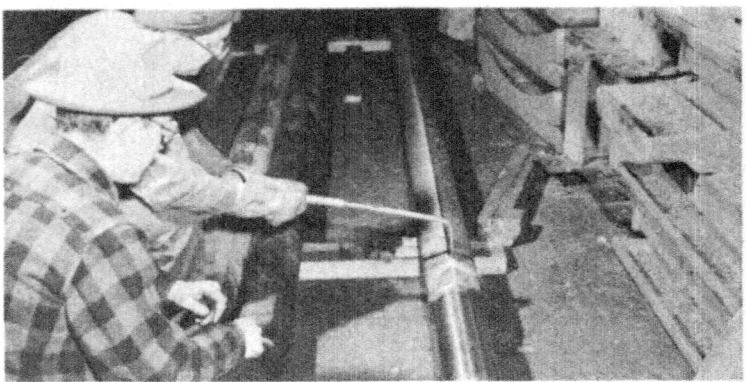

Fig. 6-3. A 5-in. shaft is straightened by spot heating.

in Fig. 6-3 is being straightened by spot heating. Some-times straightening is done by heating triangular sections, when on other occasions a rectangular patch does the trick. Experimenting is the way to find out.

Flame Hardening. By heating a piece of hardenable steel rapidly to a red heat, with the oxyacetylene flame, then quenching it is possible to obtain the maximum hardness which the steel is capable of producing, in the treated portion only, without disturbing the balance of the metal.

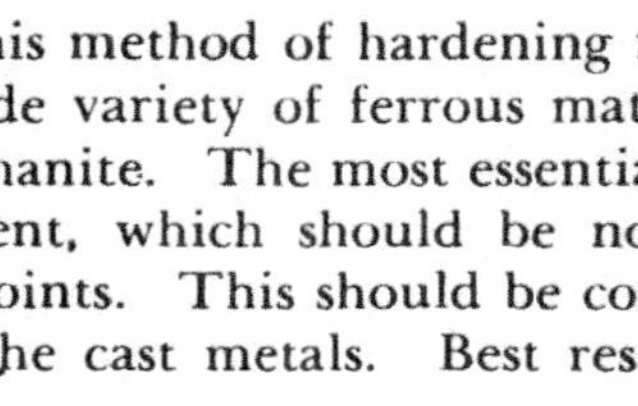

Fig. 6-4. If the member is bent away from edge B, heating should be begun at A, and continue to B along a path covering a triangular area.

This method of hardening may be used successfully on a wide variety of ferrous materials—steels, cast iron, and Meehanite. The most essential requirement is the carbon content, which should be no less than (0.35% carbon) 35 points. This should be construed as combined carbon for the cast metals. Best results are obtained when the

88

Fig. 6-5. Flat steel surfaces are flame hardened by heating and quenching as special torches move along.

carbon content is between 35 and 55 points. A greater carbon content than 55 points should be treated with care, as there is danger of cracking from overheating or too drastic quenching.

The most common reason for flame hardening is to effect a wearing surface with high abrasion-resistant qualities in combination with a tough, ductile, and shock resistant core. Flame hardening can be applied only where needed!

The hardened case itself does not stop abruptly in depth. There exists between it and the unaffected portion a buffer or transition zone in which the physical characteristics gradually changes from those of the hardened case to those of the metal in its pre-flame hardened condition. This serves to bind the case to the piece and lessen the tendency to crack off by being capable of absorbing the stresses induced by quick quenching.

Flame hardening is generally applied to the metal by means of a suitable burner arrangement followed by any of various means of quenching medium applications, such as a hose for water, a block with a drilled face for air, water,

89

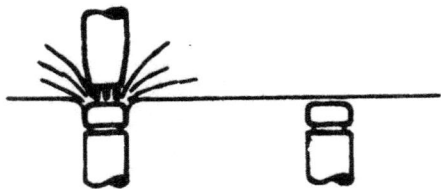

Fig. 6-6. Spot hardening of valve stems is done with a welding torch.

brine, or a soluble oil mixture, or a tank in which the work can be immersed in any of several quenching media. Usually some type of mechanism is involved to effect movement of the flame and quench along the work, or the movement of the work past a stationary flame and quench arrangement.

Spot hardening can be used for small areas, such as the ends of valve stems.   Such an operation is shown in Fig. 6-6; the valve stem is submerged but the pressure of the oxyacetylene flame blows the water away so that the end may be heated.   After being heated to temperature, the flame is removed and quenching is accomplished by the water covering the stem again.

The spinning method can be used in hardening small shafts, pulleys, rollers, or gears.   The work is mounted on a spindle and revolved, while heat is applied.

The combination method is used for hardening parts like shafts, rolls, and boring bars.   The work is mounted between centers or in a chunk, revolved at a constant rpm and the flames, followed closely by the quench, move along the longitudinal axis of the work at a rate of speed consistent with the desired result.   The method is also widely used to harden inside cylindrical surfaces, such as gear and pulley bores, or pump and cylinder liners.

Flame Cleaning. The oxyacetylene flame may be used to remove rust, scale, oil, grease, paint and other foreign materials from steel prior to welding or painting.   The equipment used depends upon the job of flame cleaning

Fig. 6-7. A special 8-in. wide heating tip is used to flame clean a bridge girder.

to be done, for some of the more simple operations a welding torch will be satisfactory while on other jobs a special flame cleaning tip may be required. Figure 6-7 is an example of the latter. In this instance a special tip 8 in. wide is used for cleaning a bridge girder.

The tip to use should be the one which is the most economical for the job.

Flame cleaning is done with a high-velocity, high-temperature flame which loosens all dirt, scale and foreign materials in such a way as to leave the surface warm and dry. The ashes left from the cleaning operation should be wire brushed from the metal. It may be painted either while warm or at a later time.

91

CHAPTER 7

Useful Data for the Gas Weldor

"He does the best he can with the tools he has." How true, no one can do a better job than the tools with which he works will let him. This applies as much, if not more so, to oxyacetylene welding than it does to any other tools with which man must work. If oxyacetyelene welding and cutting equipment is not properly cared for it will fail to do the work for which it has been designed. If care is exercised in handling this equipment it will last a lifetime with a minimum of maintenance service.

Tips on Tip Care. Some weldors seem to have trouble with the tips sticking in the torch head. Since this is a threaded connection they feel that it would be easier to change tips if they were lubricated. Remember, however, that *oxygen and oil never mix.*

To prevent sticking of the threads, lubricate them with graphite. The easiest way to do this job is to mark on the threads using a lead pencil. Merely put the point of the pencil in the thread groove, then twist the torch tip until a mark has been made the entire length of the thread. This should provide sufficient lubrication to prevent further sticking or difficulties in changing tips.

Carbon Removal. Proper care in the handling of the welding or cutting torch will avoid the necessity of much maintenance. Tips for welding and cutting torches, how-

[92]

ever, must be serviced frequently even if they receive good care.

The carbon deposited in normal welding and cutting operations may cause the tip to carbon up on the outside or the preheat holes to become fouled. Slag may also clog the cutting orifice. Any of these difficulties will impair the operation of the torch even though it has been handled with the greatest care. Wear is an additional problem as it causes a gradual enlargement or "bell mouthing" of the orifices.

The most common torch maintenance job, particularly for a cutting torch, is to clean the torch tip that has become clogged with slag. The easiest way to tackle this job is to clamp the torch in a vise with the tip pointed upward. With the torch held in proper position, it is generally easy to knock off the slag by lightly tapping the base of the tip. When the slag has been removed, the tip should be taken off the torch for further cleaning operations.

Refacing the Tip. For a bell-mouthed orifice, it will be necessary to remove all of the enlarged portion in order to return the tip to a serviceable condition. This job can best be done by rubbing the flame end of the tip on an emery cloth. The emery cloth should be placed on a smooth, flat surface; the tip should be held in a vertical position when rubbed. It is essential that the tip be held perpendicularly so that the end will be square and smooth. The tip surface should be "sanded down" until all evidence of orifice enlargement has been removed.

Cleaning Tip Orifices. When truing a tip, burrs will be formed in the preheat and cutting orifices. These must be removed with a tip-cleaning drill. Be sure you select the correct size. If in doubt as to size see Tables 7-1 and 7-2. Another precaution: be sure to work the drill up and down without any twisting motion. This will

93

Table 7-1 Drill Sizes of Various

(When cleaning tips, it is recommended that one size smaller drill be used.)

Tradename	Series	80	79	78	77	76	75	74	73	72	71	70	69
Airco	All						00			0			
Craftsman	AA									0		1	
Dockson	4EC, 4SC, 7SC			1					2				
	7EC			1					2				
Gasweld	G25, G35							00		0			1
	G55		00			0			1				2
	AVG		00			0			1				2
Harris	13, 14, 16, 17, 50 23, 13-F, 23A swedged, 17-F swedged			00			0				1		
	swedged							00				0	
K-G	EUS, KUS, KS						75			72			
Marquette	A, A1						00				0		
	B, B1			00B			0B			1B			
Meco	All	00				0						1	
Milburn	W-200	00		0		1							2
	W-100										000		
	W-600	0000			000			00					0
National	G		00			0			1				
	P					0			1				
	R						000					00	
Oxweld	W-29							1				2	
	W-17												
Prest-O-Weld	W-109						1						
	W-110, W-111						1				2		
Purox	33	0					1						
	34								1				
	35												
Rego	GP, GX, GXU, SX									72			
Smith	Lifetime										B60		
	No. 5										50		
	No. 2			18		19		A20			A21		
Torchweld	GP 570, 870									72			
	170		79			76				72			
	71					76				72			
Victor	All						000		000½			00	
Weldit	All			00			0					1	

[1] W-110 to No. 6 Tip only. [2] Airco Tip No. 13 = Drill Size No. 10; No. 14 = Drill No. 2; No. 15 = ¼ in. drill. [3] Tips No. 7 and 8 require special gooseneck. [4] 13-F and 17-F Swedged to No. 15 Tip only. [5] EUS to No. 40 Tip only. [6] KUS to No. 30 Tip only. [7] SX to No. 46 Tip only. [8] GXU and GP to No. 31 only. [9] GP 570 to No. 31 only.

Oxyacetylene Welding Tips

68	67	66	65	64	63	62	61	60	59	58	57	56	55	54	53	52	51	50	49	48
1						2						3		4			5			6
						2								3						4
		3							4				5		6					7
		3							4				5		6					7
			2						3			4	5		6					7
								4	3		5	4	5		6	7		8		
			2						3			4			5		6			7
	1					2			3			4	5		6			7		
68						62						56			53			50		
	1					2			3					4				5		
2B						3B						4B	5B			6B				7B
			2							3		4			5					
			3						4	1		5		2		6				7
00				0		2			3			4			5		3	6		4
	1							1				2	3							
		2											3		4					
		2											3		4					
		0									2		3		5					
		4									8		9	12			15			
		4									6		9	12			15		20	
2	3					4					4		6¹	5		7		6	8	
2						4					4			5			6	6		
		2									3		4			5		6		
		2													5					
68						62			58			55			53			50		
	B61						B62					B63	B64				B65			
	51						52					53	54				55			
A22			A23			A24			A25			A26	A27	A28			A29	A210		
68						62			58			55			53			50		
68						62			58			55			53			50		
68						62			58			55			53			50		
00½			0	0½			1		1½	2	2½		3	3½		4				
		2							3			4			5	6			7	

Table 7-1 Drill Sizes of Various

DRILL SIZES → Tradename	Series	47 5⁶⁄₆₄"	46	45	44	43	42	41	40	39	38	37
Airco	All			7					8			
Craftsman	AA						5					6
Dockson	4EC, 48C, 7SC				8			9				
	7EC				8							
Gasweld	G25, G35				8		9		10	11		
	G55				8							
Harris	13, 14, 16, 17, 50 23, 23A ⎫			8		9			10			
	swedged, 13-F, 17-F swedged ⎭	8				9			10			
K-G	EUS, KUS, KS			45					40⁵			
Marquette	A, A1	6				7			8			
Meco	All	6					7					
Milburn	W-200				8				9			
	W-100			5			6				7	
	W-600			7			8					
National	G		5					6				
	P		5					6		7		
	R	4				5						
Oxweld	W-29			20			30					
	W-17					30			40			
	W-26											
Prest-O-Weld	W-109	7			8							
	W-111				9				10			
Purox	33	7			8							
	34					8						
	35		7							9		
Rego	GP, GX, GXU, SX		46⁷				42					
Smith	Lifetime				B66				B67			
	No. 5				56				57			
Torchweld	GP 570, 870			46			42					
	170, 71			46								
Victor	All					5						
Weldit	All				8	9			10	11		

¹ *W-110 to No. 6 Tip only.* ² *Airco Tip No. 13 = Drill Size No. 10; No. 14 = Drill No. 2; No. 15 = ¼ in. drill.* ³ *Tips No. 7 and 8 require special gooseneck.* ⁴ *13-F and 17-F Swedged to No. 15 Tip only.* ⁵ *EUS to No. 40 Tip only.* ⁶ *KUS to No. 30 Tip only.* ⁷ *SX to No. 46 Tip only.* ⁸ *GXU and GP to No. 31 only.* ⁹ *GP 570 to No. 31 only.*

Oxyacetylene Welding Tips (Cont.)

35 36 34	33 31 32	ᵇₙ 29 30	27 28 26	25 23 24	21 22 20	19 17 18
9		10		11	12²	
	7ˢ	8				
10	11	12	13 14	15		
12 13	14 15	16				
	12	15⁴	19	22		
35		30⁶		25		20
9	10XH	11XH	12XH	14XH		
8		9	10		11	12
10	11	12 13		14	15	16
8	9 10	10	12	13	14 15	
	9					
8	9	10	9 10 11	12		
	6	7 8				
		55 70	100	90	125	150
11	12	13				
10						
	11	13		15		
36	31ˢ			25	20	
B68	B69		B610 B611	B612		
36	31⁹			25	20	
6		7 8	9 10 11	12		
12	14	16				

Table 7-2 Drill Sizes of Various

(When cleaning tips, it is recommended that one size smaller drill be used.)

DRILL SIZE → Tradename	Series	80	79 78	77 76 75	74 73 72	71 70 69
Airco	124, 144, 164 195, 198, 199					
Craftsman	B					
Dockson	All					
Gasweld	HC-31, HC-32, HC-39 WC¹-20, WC¹-35 WC¹-10, WC¹-55					00
Harris	2890-F 6290 7490-A					00
K-G	M4, M5					
Marquette	E¹ C, D¹, D1¹·³ C, D¹, D1¹·⁴					
Meco	All					
Milburn	X-100 X-2000 X-2300					
National	All					
Oxweld	CW¹-29 CW¹-23, C-31, C-32			2 2		
Prest-O-Weld	CW¹-109 CW¹-110, C-111					
Purox	33¹ 34, 35					
Rego	All					
Smith	Lifetime, Long life¹·³ Lifetime, Long life¹·⁴ Airline¹, Midline¹					
Torchweld	280, 250, 251					
Victor	All				000	
Weldit	S-25, S-37, S-45					

¹ *Cutting Attachment.* ² *Applies to M4 only.* ³ *Preheat: 4 flames.* ⁴ *Preheat: 6 flames.* ⁵ *Also oo with 3 flame preheat.* ⁶ *Applies to 280 only.* ⁷ *Tip size 11 drill size 10; tip 12 drill 2; tip 13 drill ¼ in. and tip 14 drill L.* ⁸ *Remaining tips for C only.* ⁹ *Tip 12 drill size 15.* ¹⁰ *Tip next size larger is drill size 12.*

Oxyacetylene Cutting Tips

68 67 66	65 63 64	62 61 60	59 57 58	56 55 54	53 52 $\frac{1}{16}''$	51 50 49
·00		0		1 2	3	4
		1		2	3	4
		1		2		3
0 0		1	2	1	2	
00	0	0 00	0	1 1 1 2	2	2 3
68²		62		56	53	50
	1B	2B 0A 0A		1A 1A	2A	2A
0ᵇ			1	2		
	00 00	0 0	1 1	1 2 2 2	2	3 3
		0	1	2		3
3		4 4			6	
	0	1 1			2	
0		1 1		2		3
68		62		56	53	51
	0 1	0 2		1 2 1 3 5	3 2 4	
68⁶		62		56	53	51
	00		0	1	2	3
				1	2	

99

Table 7-2 Drill Sizes of Various

Tradename	Series	48	47	46	45	44	43	42	41	40	39	38	37
Airco	124, 144, 164, 195, 198, 199				5				6				
Craftsman	B				5					6			
Dockson	All						4						
Gasweld	HC-31, HC-32, HC-39, WC-20[1], WC-35[1]				3								
Harris	2890-F				3								
Harris	6290				3								
Harris	7490-A							4					
K-G	M4, M5				45								
K-G	M6									40			
Marquette	C, D[1], D1[1,3]										3A		
Marquette	C, D[1], D1[1,4]						3A						4A[8]
Meco	All					3							
Milburn	X-100	3			4			5		6			7
Milburn	X-2000	4			5					6			7
Milburn	X-2300				4				5				
National	All			4				5					6
Oxweld	CW[1]-23, C-31			8							10		
Oxweld	C-32												
Prest-O-Weld	CW-110[1], C-111		3										
Purox	35						4						
Rego	All			46				42					
Smith	Lifetime, Long life[1,3]	4											
Smith	Lifetime, Long life[1,4]	3								4			
Torchweld	280, 250, 251			46				42					
Victor	All				4					5			
Weldit	S-25, S-37, S-45					3							

[1] Cutting Attachment. [2] Applies to M4 only. [3] Preheat: 4 flames. [4] Preheat: 6 flames. [5] Also 00 with 3 flame preheat. [6] Applies to 280 only. [7] Tip size 11 drill size 10; tip 12 drill 2; tip 13 drill ¼ in. and tip 14 drill L. [8] Remaining tips for C only. [9] Tip 12 drill size 15. [10] Tip next size larger is drill size 12.

Oxyacetylene Cutting Tips (Cont.)

35 / 36 -34	33 31 / 32	¼″ 29 / 30	28 27 / 26	25 23 / 24	21 / 22 20	19 17 / 18
7		8	9			10⁷
7						
5	6		7 8			
4 5						
4 4 / 5	5	5 / 6				
35		30		25		
		4A / 5A	6A			
4		5	6			
8 / 8 / 6	9	10 11	12			
6		7	8		9	10 11⁹
	7	8				
	12					
4 5						
35		30		25		
		5		6		
35		30		25		18¹⁰
6			7			
4		5				

101

Oxyacetylene equipment, well cared for, will produce the best of gas welds at the lowest cost. The same is true of cutting equipment. Without care, this equipment will deteriorate rapidly; without maintenance, it will soon become useless. Take care of your torch tips!

Flame Temperatures. The temperature of the oxyacetylene flame, while generally estimated to be 6,000 F, varied according to the ratio of the oxygen and acetylene mixture. Table 7—3 gives the flame temperatures for various ratios from eight-tenths parts oxygen to one part acetylene to 2.5 parts oxygen to one part acetylene.

Table 7—3 Oxyacetylene Flame Temperatures

Ratio Oxygen to Acetylene	Type of Flame	Temperature F
0.8 to 1.0	Carburizing	5550
0.9 to 1.0	Carburizing	5700
1.0 to 1.0	Neutral	5850
1.5 to 1.0	Oxidizing	6200
1.8 to 1.0	Oxidizing	6300
2.0 to 1.0	Oxidizing	6100
2.5 to 1.0	Oxidizing	6000

INDEX

INDEX

106

Printed in Great Britain
by Amazon